Praise for

REINVENTING CAPITALISM IN THE AGE OF BIG DATA

"Digitalization is challenging us to re-think the future of our economy. This thought-provoking book provides excellent insights and guidance."

—Henning Kagermann, former CEO of SAP and president of Acatech at the National Academy of Science and Engineering

"This refreshingly optimistic book, full of fascinating examples, shows how the digital age can lead to a future of data-rich markets that empower individuals and improve our lives in a diverse and inclusive society."

—Urs Gasser, professor and executive director, Berkman Klein Center for Internet & Society, Harvard University

"For a generation, information technology has progressively driven down transaction-costs and displaced a universal tradeoff between richness and reach. This landmark book takes that logic to an entirely new plane, where the richness of data merges with the open and unbounded reach of markets. The possibility of data-rich markets is a vision that should challenge and inspire every corporate strategist and public policy maker."

—Philip Evans, senior adviser at The Boston Consulting Group and BCG Fellow

"It's vogue today to proclaim the 'death of capitalism'—and yet the one truly global system is going through profound reinvention as a combination of technological forces reshape every aspect of our economic, political, and social lives. This book is an absolutely essential guide to our collective digital future and equally importantly, a sensible manifesto to shape it for everyone's benefit."

—Parag Khanna, author of *Connectography: Mapping the Future of Global Civilization* and *Technocracy in America: Rise of the Info-State*

REINVENTING
CAPITALISM
IN THE AGE OF
BIG DATA

ALSO BY VIKTOR MAYER-SCHÖNBERGER

Big Data: A Revolution That Will Transform
How We Live, Work, and Think (with Kenneth Cukier)

Learning with Big Data (with Kenneth Cukier)

Delete: The Virtue of Forgetting in the Digital Age

Governance and Information Technology: From Electronic
Government to Information Government (with David Lazer)

REINVENTING
CAPITALISM
IN THE AGE OF
BIG DATA

Viktor Mayer-Schönberger
and **Thomas Ramge**

BASIC
BOOKS
New York

Basic Books
Hachette Book Group
1290 Avenue of the Americas, New York, NY 10104
www.basicbooks.com
Printed in the United States of America

First Edition: February 2018
Published by Basic Books, an imprint of Perseus Books, LLC, a subsidiary of Hachette Book Group, Inc. The Basic Books name and logo is a trademark of the Hachette Book Group.

The publisher is not responsible for websites (or their content) that are not owned by the publisher.

Library of Congress Control Number: 2017963910
ISBNs: 978-0-465-09368-7 (hardcover), 978-0-465-09369-4 (ebook)

LSC-C

10   9   8   7   6   5   4   3   2   1

CONTENTS

-1-

REINVENTING CAPITALISM

I T SHOULD HAVE BEEN A VICTORY CELEBRATION. BY THE time eBay's new CEO, Devin Wenig, climbed the stage for the online marketplace's twentieth-anniversary event in September 2015, goods worth more than $700 billion had been traded on eBay's platform, and active eBay users had reached 160 million. The company Pierre Omidyar had started in 1995 as a small side-business turned into what looked like a perpetual money-maker. EBay had taken an old but highly successful idea, the market, and put it online.

Because eBay's market was no longer a physical place, it never closed. And thanks to the Internet's global reach, pretty much everyone connected to it could buy and sell on it. Through eBay's unique rating system, it created a way to trust market participants without knowing them. Together that made the new virtual marketplace tremendously attractive, resulting in what economists call a thick market, a market with lots of buyers and sellers. Thick markets are good markets, because they increase the likelihood of finding what one is looking for. EBay also took a feature of traditional markets and improved on it: it replaced fixed prices with an auction mechanism, a far better way to achieve optimal price, as economics students learn in their first semester.

A marketplace with global reach that's always open and makes transacting simple, easy, and efficient—that's the recipe for eBay's meteoric rise. It not only ushered in the Internet economy but also seemed to reconfirm the preeminent role markets play in our economy.

But to journalists attending the celebration, Wenig looked more like "a general rallying the troops of a beleaguered army," and his speech felt like a pep talk—with good reason. The world's largest marketplace had lost some of its mojo. Analysts on Wall Street even labeled eBay "due for a reset." With so much going for it, some may see eBay's recent troubles as a bout of bad management, aggravated by bad luck. But to us it's an indication of a much larger, structural shift.

Just months before eBay's twentieth anniversary, Yahoo, another early Internet pioneer, was suffering its own market woes. Yahoo owned a substantial chunk of Chinese online marketplace Alibaba, and based on Alibaba's share price, its holding of Alibaba's shares was more valuable than Yahoo's total market capitalization. So sellers of Yahoo's shares were essentially paying buyers to take on their stock and shares of Yahoo were trading at an effectively negative price. That doesn't make sense, of course, because the value of a share of common stock can't be negative. But stock prices, economists tell us, should reflect the collective wisdom of the market; so they ought to be right. Something was wrong—terribly wrong.

EBay's surprising troubles and Yahoo's crazy share price aren't random events. They signify a fundamental weakness of existing marketplaces, a weakness, as we'll explain, that is tied to price. Because the flaw is linked to price, not all marketplaces are suffering. In fact, some markets, less reliant on price, are outright thriving.

Just about the time eBay and Yahoo got into trouble, a more recent Internet start-up, BlaBlaCar, was doing amazingly well. Founded in Europe by a young Frenchman bitten by the Internet bug during graduate studies at Stanford, BlaBlaCar, much like eBay, operates an online marketplace, albeit a highly specialized one. It is in the business of helping people share car rides by matching those offering a ride with those looking for one. And it does so very well, matching millions of riders every month and growing quickly. Whereas eBay's original focus was on price-based auctions, BlaBlaCar's marketplace offers participants rich data about each other, ranking details such as driver chattiness (hence its name), so users can easily search and identify the best matches for them, and downplaying the importance of price (ride-sharers can select price only within a limited range). Bla-BlaCar's ride-sharing market isn't alone in using rich data. From Internet travel site Kayak to online investment company SigFig, to digital labor platform Upwork, more and more markets that use data to help participants find better matches are gaining traction and attracting attention.

In this book, we connect the dots between the difficulties faced by traditional online markets; the error of the stock market's trusted pricing mechanism; and the rise of markets rich with data. We argue that a reboot of the market fueled by data will lead to a fundamental reconfiguration of our economy, one that will be arguably as momentous as the Industrial Revolution, reinventing capitalism as we know it.

The market is a tremendously successful social innovation. It's a mechanism to help us divvy up scarce resources efficiently. That's a simple statement—with enormous impact. Markets have enabled us to feed, clothe, and house most of 8 billion humans, and to greatly improve their life expectancy as well as life quality. Market transactions have long been social interactions, making them superbly well aligned with human nature. That's why markets seem so natural to most of us and are so deeply ingrained in society's fabric. They are the building blocks of our economy.

To do their magic, markets depend on the easy flow of data, and the ability of humans to translate this data into decisions— that's how we transact on markets, where decision-making is decentralized. This is what makes markets robust and resilient, but it requires that everyone has easy access to comprehensive information about what's available. Until recently, communicating such rich information in markets was difficult and costly. So we used a workaround and condensed all of this information into a single metric: price. And we conveyed that information with the help of money.

Price and money have proved to be an ingenious stopgap to mitigate a seemingly intractable challenge, and it worked—to a degree. But as information is compressed, details and nuance get lost, leading to suboptimal transactions. If we don't fully know what is on offer or are misled by condensed information, we will choose badly. For millennia, we tolerated this inadequate solution, as no better alternative was available.

That's changing. Soon, rich data will flow through markets comprehensively, swiftly, and at low cost. We'll combine huge volumes of such data with machine learning and cutting-edge matching algorithms to create an adaptive system that can identify the best possible transaction partner on the market. It will be easy enough that we'll do this even for seemingly straightforward transactions.

Suppose, for instance, you are looking for a new frying pan. An adaptive system, residing perhaps on your smartphone, accesses your past shopping data to gather that you bought a pan for induction cooktops last time, and also that you left a so-so review of it. Parsing the review, the system understands that the pan's coating really matters to you, and that you favor a ceramic one (it also notes your preferred material for the grip). Equipped with these preferences, it then looks at online markets for optimal matches, even factoring in the carbon footprint of the delivery (because it knows how worried you are about that). It negotiates automatically with sellers, and because you are ready to pay by direct transfer it is able to get a discount. With a single tap, your transaction is complete.

It sounds seamless and simple—because it should be. It's far faster and less painful than having to do the search yourself, but it also takes into account more variations and evaluates more offers than you would do. Neither does the system tire easily (as we humans do when searching for something offline or online), nor is it distracted in its decision advice by price, derailed by cognitive bias, or lured by clever marketing. Of course, we'll still use money as a store of value, and price will still be valuable information; but no longer being focused on price broadens our perspective, yields better matches, a more efficient transaction, and, we believe, less trickery in the market.

Such decision-assistance systems based on data and machine learning will help us identify optimal matches in these data-rich markets, but we humans will retain the ultimate decision-making power and will decide how much or how little we delegate as we transact. That way we can happily have our decision-assistance system hail a ride for us, but when it comes to our next job, we'll choose ourselves from among the employment options our data-driven advisers suggest.

Conventional markets have been highly useful, but they simply can't compete with their data-driven kin. Data translates into too much of an improvement in transactions and efficiency. Data-rich markets finally deliver what markets, in theory, should always have been very good at—enabling optimal transactions—but because of informational constraints really weren't.

The benefits of this momentous change will extend to every marketplace. We'll see it in retail and travel, but also in banking and investment. Data-rich markets promise to greatly reduce the kind of irrational decision-making that led to Yahoo's crazy stock price in 2014 and to diminish bubbles and other disasters of misinformation or erroneous decision-making that afflict traditional money-based markets. We have experienced the debilitating impact of such market disasters in the recent subprime mortgage crisis and in the 2001 burst of the dot-com bubble, but also in the countless calamities that have affected money-based markets over the past centuries. The promise of data-rich markets is not that we'll eradicate these market failures completely, but that we'll be able to greatly reduce their frequency and the resulting financial devastation.

Data-rich markets will reshape all kinds of markets, from energy markets, where built-in inefficiencies have lined the pockets of large utilities and deprived households of billions in savings, to transportation and logistics, and from labor markets to health care. Even in education, we can use markets fueled by data to better match teachers, pupils, and schools. The goal is the same for all data-rich markets: to go beyond "good enough" and aim for perfection, giving us not just more bang for the buck, but more satisfaction in the choices we make, and a more sustainable future for our planet.

THE KEY DIFFERENCE BETWEEN CONVENTIONAL MARKETS and data-rich ones is the role of information flowing through them, and how it gets translated into decisions. In data-rich markets, we no longer have to condense our preferences into price and can abandon the oversimplification that was necessary because of communicative and cognitive limits. This makes it possible to pair decentralized decision-making, with its valuable qualities of robustness and resilience, with much-improved transactional efficiency. To achieve data-richness, we need to reconfigure the flow and processing of data by market participants, an idea that was already suggested as far back as 1987. Massachusetts Institute of Technology (MIT) professor Thomas Malone and his colleagues foresaw "electronic markets," but only recently have we achieved the technical progress to extend that early vision and bring it into full bloom.

One may assume that the advent of data-rich markets rests mainly on advances in data-processing capacity and network technology. After all, far more information permeates data-rich markets compared with conventional ones, and Internet bandwidth has been increasing steadily with no end in sight. Leading network technology providers such as Cisco suggest that growth rates in Internet traffic will continue to exceed 20 percent per year until at least 2021—a rate that when compounded over just a decade will add up to a staggering 500 percent upturn. Processing capacity has risen dramatically, too: we now measure our personal computer's power in thousands of billions of calculations per second, and we still have room for improvement, even if that power may no longer be doubling every two years as it has in the past.

These are necessary developments toward data-rich markets, but they aren't sufficient. What we need is to do things not just faster but to do them differently. In our data-rich future, it will matter less how fast we process information than how well and how deeply we do so. Even if we speed up the communication of price on traditional markets to milliseconds (as we have already done with high-frequency trading), we'd still be oversimplifying. Instead, we suggest that we need to put recent breakthroughs to use in three distinct areas: the standardized sharing of rich data about goods and preferences at low cost; an improved ability to identify matches along multiple dimensions; and a sophisticated yet easy-to-use way to comprehensively capture our preferences.

Just getting raw data isn't enough; we need to know what it signifies, so that we don't compare apples with oranges. With recent technical breakthroughs, we can do that far more easily than in the past. Just think of how far we have come in the ability to search our digital photos for concepts, such as people, beaches, or pets. What works for images in our photo collections can be applied to markets and can translate data into insights that inform our decision-making.

Identifying best matches is easy when we compare only by price; but as we look for matches along numerous dimensions, the process gets complex and messy, and humans easily get overwhelmed. We need smart algorithms to help us. Fortunately, here, too, substantial progress has been made in recent years. Finally, knowing exactly what we want isn't easy. We may forget an important consideration or erroneously disregard it; for humans, it's actually quite difficult to articulate our multifaceted needs in a simple, structured way. That's the third area in which recent technical advancements matter. And today, adaptive systems can learn our preferences over time as they watch what we are doing and track our decisions.

In all three of these areas, highly evolved data analytics and advanced machine learning (or "artificial intelligence," as it is often called) have fueled important progress. When combined, we have all the key building blocks of data-rich markets. Digital thought leaders and energetic online entrepreneurs are already taking note. There is a gold rush just around the corner, and it will soon be in full swing. It's a rush toward data-rich markets that deliver ample efficiency dividends to their participants and offer to the providers a sizable chunk of the total transaction volume.

The digital innovations of the last two decades are finally beginning to alter the foundations of our economy. Some companies have already set their sights on data-rich markets and put the necessary pieces in place. As eBay celebrated its twentieth anniversary and pondered its future, its new CEO announced a highly ambitious, multiyear crash program and forged a number of key acquisitions. The aim is to greatly improve the flow of rich information on the marketplace at all levels, to ease discovery of matches, and to assist eBay users in their transaction decisions.

EBay is not alone. From retail behemoth Amazon and niche players, such as BlaBlaCar, to talent markets, marketplaces are reconfiguring themselves and pushing into a data-rich future. Because data-rich markets are so much better at helping us get what we need, we'll use them a lot more than traditional markets, further fueling the shift from conventional markets to data-rich ones. But the impact of data-rich markets is far larger, the consequences far bigger.

Markets aren't just facilitating transactions. When we interact on markets, we coordinate with each other and achieve beyond our individual abilities. By reconfiguring markets and making them data rich, we shape human coordination more generally. If done well, market-driven coordination greased by rich data will allow us to meet vexing challenges and work toward sustainable solutions, from enhancing education to improving health care and addressing climate change. Gaining the ability to better coordinate human activity is a big deal.

This will have repercussions for more conventional ways of coordinating our activities. Among them, the most well known and best studied is the firm. The stories we usually tell about firms are about vicious competition between them, whether it is General Motors versus Ford, Boeing versus Airbus, CNN versus Fox News, Nike versus Adidas, Apple versus Google, or Baidu versus Tencent. We love tales about individual battles that bloodied one of the contestants and advanced the position of the other. Entire libraries of business books and hundreds of business-school cases are dedicated to chronicling and analyzing these epic battles. But rather than battles between firms, we now see a more general shift from firms to markets, as the market, thanks to data, gets so much better at what it does. This shift doesn't mean the end of the firm, but it represents its most formidable challenge in many decades.

Responding to the rise of data-rich markets isn't going to be simple. If firms could utilize the technical breakthroughs we describe, reshape the flow of information within them, and capture similar efficiency gains, it would be straightforward. Alas, as we'll explain, the technical advances that underlie and power data-richness can't be used as easily in firms as they can in markets. They are constrained by the way information flows in firms. To adapt, the nature of the firm will need to be reimagined.

Possible responses to the challenge from data-rich markets involve finding ways to either more narrowly complement or emulate them. Firms might automate decision-making of (certain) managerial decisions and introduce more marketlike features, such as decentralized information flows and transaction-matching. These strategies offer medium-term advantages, and they are being adopted in a growing number of companies. They are useful for ensuring the continuing existence of firms in the medium term (although they bring their own set of weaknesses), but they are unlikely in the long run to stop the slide of the firm's relevance in organizing human activity.

Just as firms will continue to have some, albeit diminished, role to play in the economy, in the future we'll also still use money, but in data-rich markets money will no longer play first violin. As a result, banks and other financial intermediaries will need to refocus their business models. And they are going to need to move quickly, as a new breed of data-driven financial technology companies, the so-called fintechs, are embracing data-rich markets and challenge the conventional financial services sector. It is easy to see how banking will be severely affected by the decline of money, but the implications are larger, and more profound. At least in part, the role of finance capital rests on the informational function it plays in the economy. But as data takes over from money, capital no longer provides as strong a signal of trust and confidence as it currently does, undermining the belief that capital equates with power that underlies the concept of finance capitalism. Data-richness enables us to disentangle markets and finance capital by furthering the one while depreciating the other. We are about to witness both the rather immediate reconfiguration of the banking and finance sector, and the later but more profound curbing of the role of money, shifting our economy from finance to data capitalism.

DATA-DRIVEN MARKETS OFFER SUCH COMPELLING AD-vantages over traditional, money-based ones that their advent is assured. But they are not without shortcomings of their own. The fundamental problem is the reliance on data and machine learning and the lack of diversity of data and algorithms. These make them particularly vulnerable to troubling concentration as well as systemic failure. Because of this structural weakness (which we'll explain further), data-rich markets could turn into enticing targets for ruthless companies and radical governments to not only cripple the economy but also undermine democracy. To mitigate this vulnerability, we propose an innovative regulatory measure. A progressive data-sharing mandate would ensure a comprehensive but differentiated access to feedback data and would maintain choice and diversity in decision assistance. It's not only the anti-trust measure of the data age, but it also guards against far bigger and more sinister developments that could threaten society.

The rise of a market in which a substantial part of the transactional process is automated, and the decline of the firm as the dominating organizational structure to organize human activity efficiently will uproot labor markets around the world. Nations will face the need to respond to this profound shift in the economy as it endangers many millions of jobs, fuels widespread worries in countless nations, and is already driving populist political movements. As we'll detail, many of the conventional policy measures at our disposal are unfortunately no longer effective.

A shift from finance to data capitalism will question many long-held beliefs, such as work as a standardized bundle of duties and benefits. Breaking up this bundle is going to be a challenging but necessary strategy for firms looking for the right human talent, and for societies worried about mass unemployment, to bring back to employees jobs as well as meaning and purpose. Central to the changes we'll witness in labor markets is data. Comprehensive and rich data flows drive the revival of the market and the decline of firms and money, prompting massive upheavals in the labor market. By the same token, rich data also enables us to upgrade labor markets so that they'll offer far more individualized and satisfying work far more easily and more frequently than before (although, as we explain, this will need to be supported by innovative policy measures).

From the early days of money-based markets, critics have pointed at the gap between the idea of choice, so fundamental in markets, and the actual cognitive limitations that constrain our ability to choose well. For centuries, two antagonistic views have been pitted against each other: one side has advocated for a central authority to take over decision-making in markets from vulnerable humans, while the other has defended conventional markets, and with that the concept of decentralized information flows and decision-making, arguing that crippled individual choice was far better than none. These arguments were often stark—painted in black or white.

Over the last decades, a kind of truce has taken hold around the world, an acceptance that money-based markets work, but only with the appropriate regulations in place (and with no consensus on what "appropriate" entails). The compromise is that even though we can't overcome the cognitive constraints that lead to erroneous decisions, we can put in place rules and processes that help mitigate their most negative effects. This was pragmatic, given the realities that hold sway on money-based markets, and the absence of a more enticing, workable alternative. But it was also an acceptance of defeat; real progress in improving the inner working of the market seemed forever illusive. The market was tainted, but the alternatives were worse. So, we learned to live with it.

The availability of rich data and recent technical breakthroughs mean that we now can move beyond money-based markets to data-rich ones and overcome some of the key informational and decisional constraints that we have been grappling with. The vision is ambitious. Rather than making for better mitigation of the conventional market's weaknesses, we are about to see a rewiring of the market that renders mitigation far less necessary. In the future, data-rich markets will offer individual choice without the constraints of inescapable cognitive limitations.

Of course, we won't be able to overcome all human biases and decision flaws (nor avoid savvy marketers exploiting them); even if humans choose to use smart machine learning systems on data-rich markets, that choice will still be a human one to make. When we empower ourselves to choose, we also retain that human error. Even rich data markets won't be perfect; but pragmatically, they will be far superior to what we have today. We may still err, but we'll surely err less frequently. Data-rich markets will change the role of markets and money, and question well-worn concepts, from competitiveness and employment all the way to finance capitalism itself. Because they will readjust the role of markets in coordinating human activities, they will have a huge impact on how we live and work with each other.

Some may fret over the role retained for human beings—that of the ultimate decision maker—and hope for a more rational central decision authority to take over. But we are convinced that keeping this fundamental role for humans isn't a bug; it's a feature. With the crucially important and valuable push for efficiency, sustainability, and rationality (because we really do need to improve our decision-making!), we must never forget the need to preserve and even embrace what makes us human. The ultimate goal of data-rich markets is not overall perfection but individual fulfillment, and that means celebrating the individuality, diversity, and occasional craziness that is so quintessentially human.

COMMUNICATIVE COORDINATION

T WOULD BE THE GRANDEST HUMAN PYRAMID EVER erected: a castle—or *castell*—ten tiers high, rising fifty feet or more up from its *pinya*, or base, and composed of hundreds of individuals. Other human-pyramid-building clubs in Spain's Catalonia region had attempted the feat, but none had thus far succeeded.

On November 22, 2015, the members of the Minyons club of Terrassa, Spain, tried. In front of a large crowd of spectators, while drummers and pipers played the theme of *Star Wars*, the *castellers* began to construct their castle in the air. After they'd built the ground level, the Minyons assembled a second level of ninety-six people, which would reinforce the strength of the massive tower. Above it they built a third level of forty more. On them, the rest of the more slender tower would rise or fall.

The four Minyons assigned to the fourth tier found their foothold. As the fifth-tier people locked their hands on their neighbors' shoulders, the band kicked off a traditional Catalan tune. It wasn't a premature celebration. The remaining climbers had to rely on the song's tempo to maintain their swift and highly choreographed ascent. Wincing in the unseasonably nippy wind, the crowd watched as each new foursome got into place.

Finally, it was time for the children to clamber high up into the air to crown the structure. The *enxaneta*, the climber assigned to the highest tier, had to wave to the spectators to signal that she had made it to the top before she and everyone else could carefully descend in reverse order. The moment was tense. Yes, the tower might fall apart, and the attempt would be a failure, but there was much more at stake: nine years earlier, a girl had fallen to her death from a nine-tier tower.

Nothing had been left to chance. The Minyons had started training eight months earlier, meeting twice a week, developing their strength and courage, learning the most effective ways to balance on a wobbling person's shoulders and exploring various configurations to see which one held the longest. They worked out how to tie the *faixa*, the sash worn around the waist, so that it would hold tight when climber after climber grabbed it and stepped on it like the rung in some ordinary ladder. Only after watching the group's efforts for all these months had the *cap de colla*, the head of the group, decided that they were ready to attempt the "*quatre de deu*," the four-over-ten tower. He worked with a deputy to determine the placement of people in the base and bottom tiers to ensure an even distribution of support to all four sides of the tower. The pyramid would only be considered "complete" if it did not collapse as it was deconstructed, which meant that the bottom tiers had to hold firm for nearly four minutes as the weight of people constantly shifted above them. When the Minyons completed their tower, they had built their human castle and set a new world record. As a result of their diligent coordination, it seemed as though there were "no limits but the sky."

For the Catalan people, building *castells* is a tradition stretching back three hundred years to the convention of creating a small human pyramid at the end of a popular folk dance. How this practice evolved into building *castells* with hundreds of people isn't quite clear, beyond our quintessentially human impulse to reach a goal—then another and another until we reach for the stars. No one gets paid to be a *casteller*; money has nothing to do with it. There are, however, several points of pride on the line.

Castell competitions are held every two years. The "winner" is not always the team that has built the tallest tower: the complexity of the structure is the principal concern, as it reflects the degree of human coordination involved. An eleven-story tower with a single person on each tier is a much simpler accomplishment, requiring far fewer people, than a ten-story tower comprising three or four people on each tier. The more people involved, the more astonishing the spectacle. Because so much depends on coordinating from the bottom to the top, *fer pinya*—Catalan for "to make the base"—has come to mean "working together" generally.

The *castells* of Catalonia are a remarkable example of human coordination. The tower building requires significant preparation, including copious amounts of time and effort to observe what works and imagine what might yet be possible. Most important, it demands faultless communication. The head of the club shouts guidance from the ground, but that cannot be the only information conveyed up and down the *castell* as it is erected. Climbers must constantly communicate their standing in the tower, letting the people beside them know if they are starting to struggle under the weight or lose their balance. Information flows through gestures as well as words—a squeeze of a shoulder or the trembling of a foot are important clues about the potential for success or the imminent danger of failure. The team's members must respond to the information dexterously, as too great a shift by one person can push others out of alignment and trigger a collapse. An adjustment here or there can save the structure; at the very least it will ensure that everybody falls safely into the many arms that make up the roof of the *pinya*. A delicate give-and-take is essential to achieving the goal, as has been the case for generations of *castellers*.

Despite the importance of the moments when humans first tamed fire, invented the wheel, or developed the steam engine, these discoveries and inventions pale compared to our human ability to coordinate. Without coordination, a flame would not warm more than one human being; the wheel could not transport but a single individual; and the steam engine would have no tracks to roll on and no factory to operate in. If there is a single crucial thread that has persisted through human history, it is the importance of coordination, whether our aim is to build a *castell* or a country. Close coordination played a transformative role in human evolution; in fact, our very existence has depended on it. Although early hominids were learning to stand upright, they remained easy prey for the big predators stalking the African savannas. Only when they came together, shouting alarms and fashioning tools and reshaping the world to their benefit, could they improve their living conditions. Coordination allowed our ancestors to combine their strengths, and as a result they lived longer and thrived, generation after generation. By forming familial bonds and banding together, it became possible to protect a dependent child for several years after birth, giving humans time to develop and nurture extraordinary cognitive capacities and skills.

As humans grew ever more proficient at large-scale coordination, they were able to accomplish far more than generations before them. Coordination enabled the design and construction of breathtaking monuments, from the pyramids of Giza, the Mayan temple of Chichén Itzá, and the sprawling Angkor Wat to St. Peter's Basilica and the Taj Mahal. Their complexity and sheer scale display our amazing ability to bring people together, in labor as well as worship, devotion, and love. Other feats of engineering that seemingly served more prosaic purposes also defined who could coordinate with whom. The Great Wall separated the Chinese empire from the encroaching Mongol hordes and kept a lid on centuries of Chinese technological advances in metallurgy and agriculture. When the Suez Canal opened in 1869, it cut the sea route from Europe to Asia by 30 percent and opened the floodgates to globalization.

The monuments to our power to coordinate are not limited to large physical structures. The library of Alexandria and its hundreds of thousands of scrolls, too, was a testament to human coordination, as it pooled the knowledge of the ancient world—it is said, by forcing visiting merchants to surrender their original books in exchange for a freshly transcribed copy. The revolutionary eighteenth-century *Encyclopédie* was a joint effort among many dozens of France's greatest intellectuals, who gathered everything that they believed an enlightened citizenry needed to know into 71,818 articles, free from the stranglehold of a dictating authority (the Jesuits). Indeed, *Wikipedia*'s ability to effectively and efficiently coordinate hundreds of thousands of contributors to create more than 40 million articles in nearly three hundred languages is just the latest in a long line of collaborative projects aimed at capturing our understanding of the whole wide web of the world.

Even the pinnacles of scientific achievement—many of which we ascribe to a single mind—are often the product of coordination. Carolus Linnaeus may be credited with inventing the first taxonomic system to classify the planet's life forms, but he depended on an extensive network of patrons, colleagues, and students to collect samples far from his native Sweden and its limited biodiversity. Without their help in creating this vast catalog, Linnaeus could not have made his argument that each species had unique characteristics and an "allotted place" in nature—concepts that directly led to the theory of evolution. The moon landing required not just one Neil Armstrong stepping into the powdery lunar dust or the staff at the National Aeronautics and Space Administration (NASA) mission control center commanding the launch of the Apollo spacecraft. It also required more than 300,000 mathematicians, physicists, biologists, chemists, engineers, and mechanics spread across dozens of labs, each playing his or her own small part, from developing a menu of foods to sustain people in zero gravity to setting up a communications link between the lunar module, mission control, and the White House to crafting the parachute that safely brought the astronauts home to the blue marble of Earth. Similarly, the construction of the Large Hadron Collider, which in 2012 detected the Higgs boson and helped solidify the Standard Model of particle physics, involved more than 10,000 scientists from over one hundred countries. We do not unravel the mysteries of our universe and our existence through the work of a single lone genius but rather through collaboration among many other individuals. As one of Linnaeus's students put it, "He who holds the chain of things looks with grace upon each link."

The varieties of human coordination are as diverse as human populations, from the web of reciprocal responsibilities and duties within a social network of family and kinship to the centralized command and control of an army to the collaborative peer production of encyclopedic projects and scientific experiments. "Coordination ranges from tyrannical to democratic," wrote Yale economist Charles Lindblom. "My notion of a well-coordinated or organized society might envision a dominating elite—Plato's philosopher-kings or an aristocracy, for example. Yours might envision egalitarian institutions."

Human coordination rests on our faculty for communication. We acquired and developed complex languages to convey nuances and to enlist other individuals to help us reach our goals. We negotiate and forge partnerships through conversations, correspondence, and contracts. With the written word, we gained a tool for transferring information through space and time, giving us the means to express ourselves across miles and into the future.

Advancements in the flow of information often underlie a step-change in our coordinative capacity. Assyrian cuneiform enabled our ancestors to organize by recording harvests and transactions. Ships would not only return with precious wares from distant lands, they would also bring back information for armies and merchants. The invention of the telegraph, telephone, and other communications technologies—including the Internet—have greatly improved human coordination through effective communication. And societal institutions help humans coordinate through subtle communication: courts, for instance, send signals about how specific conflicts are settled, thereby reducing the incidences of future disagreements. In their own unique way, all these different ways of communicating influence our ability to coordinate.

Some tools of communication turn out to be better suited for a particular kind of coordination than others. For example, written notes take time to reach the recipient and require both sides to be literate and share the same language, but they can be very precise and detailed. A foreman at a factory floor can holler commands to a group of workers and thereby share information swiftly with a number of others, but there is a limit to how easily information can flow back. Similarly with mobile phones, it's easy to reach someone with a phone (if there is network coverage), and the spoken give-and-take is more flexible and faster than written communication, but it's harder to coordinate an entire group of people that way. Changes in how we communicate have had a profound impact on the way we coordinate.

THE MOST OBVIOUS WAY TO MEASURE SUCCESS OF OUR coordinated and cooperative efforts is in terms of *effectiveness*. Did we win the battle? Did we seat the capstone? Did we catalog all that is known about astronomy? Did we part the waters? Did we put a man on the moon? Effectiveness is about the ends, not the means: it's about achieving the result, no matter the cost.

The pharaohs of ancient Egypt did not worry much about the cost of building the pyramids, nor did Emperor Qin when he led his army in the conquest of the Yue and Xiongnu tribes, expanding the Chinese state and building the first "long wall" to defend it. These leaders, and those following in their footsteps, were far more concerned with effecting their visions than with the price tag for doing so. Likewise, a community may decide to harvest a crop from a plot of land, even if this wastes a great deal of water. An army may want to win a war, even at the expense of a great number of soldiers. It doesn't matter if it cost $10 billion to build the Large Hadron Collider, the scientists suggest, because the knowledge gained from it is priceless—it will lead us to innumerable other discoveries (but policy makers still worried about the cost).

The truth, of course, is that our resources aren't unlimited. Only in paradise do milk and honey flow aplenty. Throughout the ages, resources have been scarce, and our means for utilizing them have been limited. Thus, for most of us, in most circumstances, it was never enough to simply reach a goal irrespective of cost; we had to accomplish our aims *efficiently*, avoiding waste. The very origin of the word *economics*—the Greek *oikonomia*, or "rules of the house"—refers to the ancient practice of managing an estate with self-sufficiency and frugality. In the early twenty-first century, with more than 7.5 billion people to feed, clothe, house, educate, and employ in the world, we are facing numerous constraints on crucial resources—not just natural resources but also those of money and time. More than ever we strive to coordinate efficiently through improved communication.

There are two mechanisms that have been absolutely critical in helping us coordinate successfully at scale. These amazing social innovations not only make it easy for humans to work together but also ensure that they do so efficiently. With them, we have been able to accommodate fast global population growth and breathtaking increases in life expectancy: just within the last five hundred years, the number of people inhabiting the world has grown almost twentyfold, and life expectancy has almost tripled. Accommodating so many humans, their needs and desires and their hopes and dreams, necessitates coordination mechanisms that are amazingly effective and astoundingly efficient. These two innovations represent a huge advance in our efforts to coordinate, and we have rightfully embraced them in countless settings and in most societies around the world. Both are so familiar that we often take them for granted, but they are crucial to what we have achieved. They are the market and the firm.

But while they aim to achieve the same thing—helping humans to coordinate efficiently—they do so very differently. One of the decisive differences is in the way that information flows and decisions are reached. In a market, coordination is decentralized. Individuals in the market gather and provide information and make decisions for themselves. In a competitive, well-functioning market, there is no single leader deciding what is being bought or sold and under what conditions, no central authority that tells people what to do and when to do it. Because coordination is diffused, markets are flexible and dynamic. Adding participants is easy. People can join or leave the market at will. As a population grows, the market grows with it; as people travel and communicate over increasingly long distances, the market encompasses outsiders and newcomers. As Charles Lindblom observed, through the market, coordination is possible not just on the level of a household or village but also on the level of a great city or society—without having to depend on just a handful of people to anticipate (or try to anticipate) everyone's wants and needs. In other words, the market scales extremely well.

Market coordination takes place through transactions, when buyer and seller discover they have matching preferences and agree on the terms of a deal. Myriad transactions take place in markets around the globe every day. Each of us engages in dozens of them every week, from the coffee-to-go in the morning to purchasing a new dress at the mall or taking a date out to dinner. Globally, transactions worth well over $100 trillion take place every year, a figure that has grown by a factor of almost 2,000 since the 1500s. And every such transaction boils down to two parties communicating with each other. It's an amazing feat—all achieved through a simple social innovation. The great Scottish philosopher Adam Smith coined the term "the invisible hand" to capture the essence of what makes markets work, nearly 250 years ago. But the simplicity of the metaphor conceals a complex and astonishing accomplishment that altered the conditions for coordination. It has to do with how much our goals have to be aligned for human coordination to happen.

In many instances, when humans work together toward a common goal, they must share that goal. One party needs to induce, cajole, persuade, and prod others to set aside personal priorities and preferences, if only temporarily. Where it works, it enables many to work together effectively, but keeping everyone on the same page for long is difficult, and joint efforts regularly fail. In the absence of persuasion, humans have sometimes resorted to cooperation based on coercion, not on choice. Even if that succeeds, it is neither morally just nor, as many coercive regimes have learned, particularly durable.

In contrast, the market does not require participants to share their individual goals for transactions to take place, nor is it based on compulsion; instead, participants are permitted, even encouraged, to further their own immediate interests by accepting only those transactions that they find personally advantageous. This process greases the machinery of human cooperation to everyone's benefit.

The market is not the only social mechanism to enable coordination. It shares the limelight with the firm. Even though we often think of a firm as part of a market system, the truth is that the market and the firm adopt complementary and contrasting approaches to the problem of efficiently coordinating human activity. In essence, market and firm are rivals for our coordinative capacity.

The firm is no less successful in helping individuals coordinate with each other. In most countries, well over two-thirds of the workforce is employed by the estimated 100–200 million firms that exist around the world. Over the last decades, the share of people working in the private sector in many nations has grown, especially as employment by private-sector firms in high-growth countries such as China has skyrocketed. In the developed Organization for Economic Cooperation and Development (OECD) nations, almost four out of five humans work in a firm. These firms can be tiny, employing only a handful of individuals, or gigantic, like the US discount retailer Walmart, which employs more than 2 million people, or anything in between.

However, the firm—unlike the market—is an example of centralized coordination, featuring an equally centralized communicative structure. People come together in a firm to pool their efforts and resources, but their activities are organized and directed by a single recognized central authority. There is a relatively stable group of members, with participants clearly inside the firm for a period of time. Outsiders must be carefully vetted; newcomers must be thoroughly oriented. Individuals with relevant experience are charged with making key decisions with a specific goal in mind—typically, though not always, maximizing the firm's profits for its owners and shareholders. Leaders may have expertise related to the firm's competitive advantage or because they are good at motivating employees and persuading customers. Each member of the firm is given a clear set of responsibilities, and people are usually brought into the firm because their skills fit a stated strategy. Because of the division of labor, decision-making in most firms is hierarchical and centralized.

Henry Ford was a famous devotee of hierarchical, command-and-control management. When the first prototype of the Model T rolled off the factory floor, on October 1, 1908, the market for cars was just emerging. Ford's success was related less to the design of the cars than to his control of the manufacturing process. Instead of having workers move from one car on the shop floor to the next, he had the workers remain stationary and brought the cars to them on a series of moving assembly lines. This and many other innovations cut the amount of time it took to produce a car by more than half. To solve the problem of the length of time needed for the car's paint to dry, Ford used his own special recipe for japan black, a lacquer that dried in forty-eight hours, much faster than any other formula or color he tested. Ford's approach to production slashed the price tag of one of his company's cars to an affordable $825 when it was introduced in the market in 1909; by the mid-1920s, Ford's Model T sold for less than $300.

Ford maintained strict rules, both on the factory floor and in his workers' homes. When high employee turnover was threatening efficiency, he increased wages, implementing the "five-dollar day"—but the rate was only granted to those who met the standards of Ford's "sociological department," which gathered details about the character of employees and monitored their drinking, spending, and even their household tidiness.

Ford did not want to share decision-making authority with anyone. When the firm's shareholders demanded a larger dividend, he borrowed money not just to pay the dividend but also to buy back the company, putting it under his sole control.

When sales slumped in 1920, he shut down his manufacturing units for nearly six weeks and eliminated anything he viewed as waste, including 60 percent of the company's telephone lines. By his reckoning, "only a comparatively few men in any organization need telephones." After all, important information should flow upward—to him, in the head office—not laterally. The following year sales doubled, while prices fell. The company was back on track.

Many firms, and not just those in the automotive industry, have followed Ford's model of combining the division of labor with centralized decision-making. These companies manufacture products within a tightly controlled, largely vertically integrated organization. Some critics of capitalism have argued that firms will increase in size and combine to form monopolies or oligopolies that may ultimately control the economy and undo the market as we know it. Although we have seen vast concentration in a number of sectors—from trains and steel in the late 1890s to huge conglomerates (sometimes called national champions) in the latter half of the twentieth century to digital behemoths such as Amazon, Google, Facebook, and Baidu in the twenty-first century—the firm hasn't yet replaced the market. Firms and markets still compete against each other to predominate when efficiency matters. And in some sectors such as manufacturing, which were once dominated by firms, a shift is underway to organize through the market.

For example, in the 1990s a number of state-owned companies in China teamed up with the "big four" Japanese manufacturers (Honda, Kawasaki, Suzuki, and Yamaha) to build motorcycles for the growing Chinese domestic market. The Chinese companies licensed the designs from the Japanese developers and, like the Ford Motor Company, built each part to exacting specifications. But at around $700, despite being much cheaper than the equivalent models manufactured in Japan, these motorcycles were still well beyond the budget of most Chinese citizens. According to researchers John Seely Brown and John Hagel, after the government opened up the industry to small entrepreneurs, several companies clustered in Chongqing Province broke away from the licensing system in an effort to create a less expensive process that would make motorcycles affordable to the masses. Instead of looking for ways to decrease the expenses in their own factories, these companies decided to buy and assemble parts made by others. They went to the market.

First, the assemblers broke down the design of the most popular motorcycle model into four basic modules, each made up of hundreds of components. They then distributed sketches of these modules to every possible parts supplier, leaving almost all the details out. Potential suppliers had to ensure that their parts met basic standards for weight and size and worked seamlessly with the other parts in the module. Beyond that, they could make any improvements in the design they wanted to, especially if they reduced the cost—to themselves, to the assemblers, and to consumers. The assemblers didn't dictate anything. Perhaps most un-corporate of all was the fact that there were plenty of decision makers in the manufacturing process—all of them on equal footing.

Many of the assemblers also made it clear that they were not going to enter into exclusive contracts with any one supplier. That would be too constraining. They wanted the freedom to buy the same or similar components and modules from multiple sources, to be able to switch and swap based on availability and demand, and to respond to new information about the features consumers found most appealing. With millions of interchangeable parts being churned out in Chongqing, even small "mom-and-pop" shops could get into motorcycle assembling, dramatically expanding the number of market participants.

Using this modular, market-based production process, the price of a motorcycle plummeted to under $200. By 2005, Chinese manufacturers accounted for half of the global production of motorcycles, and in several emerging markets, they overtook Japanese brand names. Honda's sales fell from 90 percent to 30 percent of the market in Vietnam within only five years. The Chinese had not only deconstructed the basic architecture of state-of-the-art Japanese motorcycles, they had also deconstructed the basic organizational architecture of motorcycle manufacturing. Rather than opting for a firm's centralized control and vertical integration, they succeeded by drawing on participants in a market to efficiently produce affordable motorcycles.

Decentralized and diffuse or centralized and hierarchical? This is the choice we face when we want to coordinate efficiently. Do we opt for the market or choose the firm? Each offers unique qualities, and each differs starkly from the other. As much as they are complementary at times, there is no question that markets and firms are two distinct social innovations, two powerful mechanisms that help humans coordinate with each other, two amazing strategies competing fiercely with each other.

The key difference between the market and the firm is in the way information flows and is translated into decisions, and by whom. This is reflected in their structures: the market mirrors the flow of information from everyone to anyone and the decentralized decision-making by all market participants, much as the hierarchical firm mirrors information streaming to its center, where leaders make the key decisions. Of course, not all car manufacturers work like the Ford Motor Company, and not all markets exactly resemble the one for motorcycle parts in Chongqing. Diverse contexts have produced a variety of well-functioning structures in firms and markets.

More important, at different times the market has had a competitive advantage over the firm, and vice versa. Since the beginning of the nineteenth century, and propelled by new methods and tools that have advantaged the firm's specific structures for information flow and decision-making, the firm has risen dramatically in importance.

This advantage, we suggest, is not only temporary, it is already coming to an end. The data age has introduced an unprecedented counterforce that will push the market forward, opening not only a new chapter in the age-old competition between market and firm, but also offering society a vastly more efficient way to co-ordinate its activities. To appreciate how this has been possible, we need to first understand the information flows and decision-making processes in traditional markets.

−3−

MARKETS AND MONEY

I N THE EARLY MORNING HOURS DURING FISHING SEA-
son, hundreds of boats push out from the towns and villages
of the state of Kerala, on India's Malabar Coast. Because the fish
they catch—primarily sardines and mackerel, mainstays of the
local diet—must be sold and used relatively soon after being
brought to shore, numerous markets have sprung up in villages
along the coastline.

For hundreds of years, Kerala fishermen were confronted
with two basic choices when it came to selling their fish. On a
particularly successful day, when a fisherman pulled in a great
haul, he would have no idea whether other boats working in the
area were having just as much luck as he had, but he would know
there was a chance of it. This forced him to make a risky calculation: he could steer his boat to the closest market, which would

cost him the least amount of time and energy. But when he got there, he might find himself competing with many fishermen and get little in return for his day's work. There was even the possibility that by the time a fisherman landed his boat, the local demand would have been fully satisfied. Then he'd get nothing at all.

Alternatively, the fisherman could gamble and land his boat farther down the coast, incurring a greater expenditure of time and fuel. However, if other fishermen were making the same calculation, there was no guarantee that the distant market would be any better than the close one. And once he chose his market, he was basically stuck with it. His fish could very well spoil during the time it would take to travel up and down the coast looking for buyers. Thus, if a fisherman couldn't sell his catch at the market where he'd landed, he would usually just throw it away.

Yet often, as it turns out, there were buyers nearby—less than ten miles away, in some cases—who weren't able to get fish and were willing to pay a premium for it. The fishermen just didn't know it. Neither did the buyers on land know how much fish would be available. Their only choice was to trust what was already on offer. As a result, prices for fish were incredibly volatile, with wild swings in each local marketplace—an indication of huge inefficiencies in the market overall.

digital ✕Then, in 1997, mobile phone towers were installed in a series of coastal towns, extending reception well into the sardine and mackerel grounds offshore. Soon, as Robert Jensen, a professor at the University of Pennsylvania's Wharton School, has explained eloquently, the fishermen were transacting with buyers while they were still out on the water. As information about the supply and demand for fish in various markets got distributed more widely, market volatility plummeted. Thanks to a better flow of information, the market became vastly more efficient.

The story of the Kerala fishermen adopting mobile phones has been described as a case of empowerment through digital technologies, and as a compelling demonstration of the importance of information to the success of a market. For us, however, though correct, these characterizations miss a crucial point: not every digital technology empowers market participants, nor will an additional information flow necessarily improve markets. Whether a particular technology furthers the market by enabling new information streams depends on how well the specific qualities of that technology are aligned with the informational structure of the market.

For the Kerala fishermen, mobile phones were such an empowering communication tool because they enabled one-on-one conversations with their potential buyers. This led to more and better transactions, greatly improving the working of the market. In contrast providing fishermen with a gigantic megaphone to advertise their catch to the markets on shore would not have helped much, as information would have flown in one direction only. And if everyone had a megaphone it would have been nearly impossible for a fisherman to communicate with any one buyer. With mobile phones, information about product and price—the crucial pieces of information needed in conventional markets—could be exchanged swiftly. Communication was efficient and timely. The secret of success was the excellent fit between what mobile phones enabled and the kind of information flows—simple, fast, two-way, and across distance—the market needed.

IN THIS CHAPTER, WE WILL EXAMINE HOW THE STRUC-
ture of the market is linked to information and how that information flows, how it is translated into transaction decisions, and
how the information role of money has been decisive in making
traditional markets successful—up to a point.

The fundamental principle of the market is that decision-
making is decentralized, and so is the flow of information. People evaluate the information available to them and use it to make
decisions that benefit them. Information flows *from* everyone *to*
everyone.

Of course no one in the market can know everything—but
the market does not require omniscience. When participants
learn new information, it influences their priorities and preferences, which in turn are reflected in the choice of transactions
they engage in as well as those they forgo. For example, if a vendor in a farmer's market routinely proffers bad apples, buyers
will choose to patronize a different stall the next time they want
to buy fruit. Shorter lines in front of that vendor's stall signal the
decision of some buyers to purchase their apples elsewhere. Customers don't have to try the apples at every stall to get a sense
of each vendor's standard of quality; they can use the length of
customer lines as a proxy. It's not perfect, but it's a good and
quick first approximation. Information leads to efficiency gains,
not just for the market as a whole but also for individual participants. It beats having to investigate every potential transaction
partner in the market by yourself.

Decentralized decision-making helped by a wealth of information has another important advantage: it mitigates the effect
of bad decisions. When a central authority is making a decision
for everyone, a lot depends on the authority getting this decision
right. In the market, on the other hand, the consequences of a

single bad decision are comparatively local. If one person makes a wrong choice, the market as a whole does not collapse; there is no single point of failure. This makes the market quite resilient. And the bigger the market and the more diverse its participants, the more resilient it becomes. Once an individual discovers she made the wrong decision, that will likely be factored into her future decisions, which in turn sends signals to the market. Because of such informative signals, not just the individual but also the market learns—not in a controlled, linear, or clearly predictable fashion, but it learns nonetheless.

Occasionally far more than a few people make the same mistake, and the market suffers. Cascades of bad information can lead to bubbles and sudden crashes. But in markets that are working well, these systemic blunders are rare relative to the volume of transactions. In the words of economist and Nobel laureate Friedrich August von Hayek, "The market is essentially an ordering mechanism, growing up without anybody wholly understanding it, that enables us to utilize widely dispersed information about the significance of circumstances of which we are mostly ignorant."

There is a critical link between market efficiency and information flows, and the experience of the Kerala fishermen is a powerful illustration. Information can make or break a market—not only does information have to travel throughout the market, but it must travel at low cost. Each additional ounce of effort and each extra penny expended in the pursuit of necessary information makes the market a more expensive mechanism for human coordination. Little would have changed for the fishermen in Kerala, for example, if placing a call from their mobile phones had cost more than they earned for a day's catch or if technical problems had forced them to dial dozens of times before they could get through. And every additional cost that turns into a reason for market participants to *not* pursue a piece of information increases the number of bad decisions.

Of course it is only in ideal markets that each participant always has all the information she needs. The reality is more challenging. Some participants, for example, may not reveal their preferences openly, in order to strengthen their negotiating positions and force a better deal. This may sound like a sensible strategy for an individual, but if it is widespread, it hurts everyone by making it difficult for others to process the information being shared. Moreover, if market participants have to assume that others are not transparent, they must factor this into their decision-making. In his classic example of information asymmetry, George Akerlof cites the market for used cars. Because it's difficult to inspect the condition of every component of a car without disassembling it, buyers cannot really ascertain if a car they are considering purchasing is a "peach" or a "lemon" at the time of the transaction. As every used car in the market could potentially be a lemon, buyers are less inclined to pay extra for a purported peach, while sellers who actually have a car in great condition must absorb the market's informational inefficiencies, and in most cases they either decide not to sell their cars or to sell them for less than they feel they're worth. As a result, fewer peaches are offered for sale, which reduces buyers' options in the market. This "lemon problem" highlights how a lack of information in the market leads to a decision-making dynamic that hurts not just individual participants but the market as a whole.

There is still some disagreement among economists concerning how much information an efficient market requires. As we have seen, if there is too little information, bad decisions will result. But the reverse can pose problems, too: in a market where everyone knows everything about everyone else, participants with new ideas may not be able to profit enough from them before copycats appear and free ride (hence the perceived need for intellectual property protection). And if everything is communicated to everyone, the sheer volume of information might be too difficult and costly to process.

Still, the overwhelming view among economists is that in markets, more information trumps less. This is why rules mandate the sharing of information in many markets. In the United States, for example, people selling their cars are required to inform buyers of any major accidents the car has been involved in. Companies listed on the stock market are required to file quarterly financial reports with the stock market regulator, which are then made public. Banks and investment funds, too, must comply with stringent reporting obligations (although, as we have seen in the subprime mortgage crisis, if they bury pertinent information deeply enough, potential investors may not notice). In many jurisdictions, doing business directly with consumers obliges the seller to fully disclose any unusual contractual terms before concluding a transaction. And companies operating in certain sectors, from pharmaceuticals and health care to education and air travel, are required to provide additional information to regulators and the public.

Even when an individual doesn't intentionally withhold information, there can be obstacles to its free flow. When a piece of art sells at a flea market price and then turns out to be a valuable original, the information about the true value of the goods has somehow gotten lost, and a transaction takes place that shouldn't. In such cases, one side suffers a financial loss. Such informational failures can lead to more tragic consequences, when important—even life-saving—insights are available to a limited number of people but do not spread fast enough to reach the people who urgently need them.

Consider the case of Vicki Mason, a young British woman pregnant with her first child in the autumn of 1961. To counter her morning sickness, she took a new sedative from a German pharmaceuticals company, Grünenthal, which had a flawless reputation. It had been suggested by her doctor, and it appeared to be so risk free that the British government was allowing a subsidiary of the beverage company Distillers to distribute it over the counter. By the time Vicki started taking the new drug—also known by its generic name, thalidomide—a German doctor, alarmed by the growing number of babies being born with misshapen limbs, had started actively investigating the connection to the drug's use. By mid-November, he informed Grünenthal of his findings and by the end of the year, thalidomide was no longer available for sale in West Germany or the United Kingdom. Vicki's daughter Louise, born in June 1962, was the last British "thalidomide baby" to survive beyond infancy. Vicki Mason had no way of knowing that she was making a horrific mistake when she decided to take Grünenthal's drug. Data on the side effects had not reached her or her doctor in time. Eventually, important information may spread to all corners of a market, but if it isn't available in time for those facing a decision, it may lead to grave errors.

LACK OF INFORMATION, HOWEVER, ISN'T THE ONLY challenge. For decades, economists have presumed that transactions are the outgrowth of rational calculations. If a person prefers bananas over apples, for example, and is offered both at the same price, she will choose to buy bananas. Decisions were seen as the logical consequences of a person's preferences and constraints—of what was demanded and what could be supplied. As it turns out, however, market participants make far more bad decisions than one would expect. Sometimes this is engineered by marketing tactics. When shopping for groceries we buy more when shopping carts are bigger. We buy more cheese than we actually need after a charming salesperson offers us a few bites to taste. And many of us give in to temptation and buy the candies, gums, and magazines out of boredom while waiting on the checkout line. Our transaction decisions are clouded by human irrationality.

Even if we aren't exposed to any such persuasive marketing efforts, we can become overwhelmed by the complex task of matching our preferences with what is available on the market. Suppose we prefer bananas over apples, but also organic over conventional and ripe over green. How would we choose between green conventional bananas and ripe organic apples? It isn't a simple matter of weighing the pros and cons for each choice: we also have to weigh them according to their importance. Quickly we'll face a pretty vexing decision. Although knowing more about our preferences and options is helpful in general, having to actually weigh, factor, and compare this information in all its dimensions (not just the type of fruit but also its ripeness, how and where it was grown, and perhaps its sugar content, nutritional value, and shelf life) may overwhelm our mental capacities and lead us to make decisions that aren't entirely rational. It may not matter that much when we choose a fruit at the supermarket. But it matters quite a bit when we are faced with more consequential choices: what hotel we book for our annual vacation, which new car we purchase, what house we get, which school we pick for our children, or what medical treatment we choose may hinge on our successful processing of many different dimensions of preferences.

Sometimes sellers deliberately make it hard to assess and compare the products and services available by adding even more dimensions or by providing information for each dimension in nonstandard form. Think of insurance policies. Deciding well in these circumstances is hard. Except when it comes to recognizing visual patterns, the human brain isn't very good at processing huge amounts of information. In experiments, psychologists have found that humans are only really able to juggle about half a dozen distinct pieces of information at the same time—not even enough to be able to compare and contrast three characteristics of three different products.

It's a frustrating conundrum: on the one hand, we yearn for more information to assess our options and transact wisely; on the other hand, we are being overwhelmed by information, fail to process it successfully, and risk making a less than optimal choice. We may not like it, but in such situations, sometimes we find ourselves stymied: either we know too little and thus can't recognize the most appropriate choice, or we know so much that, overwhelmed, we choose poorly.

The excessive cost of information and our limited capacity to process it often lead us to make mistakes. Yes, if we know and try hard, we may control our temptation to buy those supermarket candies. But we can't as easily overcome the limitations that are hardwired into our brains when comparing multiple items along multiple dimensions. This limits our ability to make the most of markets. Even if we discover an inexpensive, fast way to communicate relevant information, we are still restricted by our cognitive abilities; and even if we augment our cognitive capabilities, it is not enough if information does not reach us or does so too slowly or at too high a cost.

Yet as intractable as this challenge may sound, a fix is available that mitigates these problems, and we have been using it for millennia: money.

"MONEY IS THE ROOT OF MOST PROGRESS," HARVARD historian Niall Ferguson wrote in his widely acclaimed work *The Ascent of Money*. Money's importance is directly linked to its utility. Its obvious role is that it stores and holds value. When trade was transacted with gold and silver coins, this seemed self-evident; precious metals are rare, so coins made from them are valuable. But money has another role. With money, we can condense information about our preferences into price and this information can be conveyed and processed by humans much more easily.

Using money and price we make markets work. Money acts as a standardized yardstick to denominate the value of goods and services, allowing people to size up dissimilar items, to compare apples to oranges, coffee mugs to teacups. In the absence of money, when individuals bartered in the market, they had to come to some agreement about how much of one good should be exchanged for how much of another. That was terribly difficult without an accepted common denominator. It created unpredictability and made it difficult to correlate transactions. Knowing that an individual traded a knife for a fur coat is not much help to someone wanting to trade a slab of reindeer meat for a vessel full of fish oil. Bartering provides little information to anyone who isn't trading in exactly the same entities. With money as an accepted yardstick, however, negotiating transactions not only got easier, but the information generated from such transactions could be shared. Through money and price, transactional information got a standardized language that market participants understood. Goods and transaction partners varied, but the informational value of each transaction persisted in an easy-to-understand vernacular, to inform and enlighten the market.

This offers yet another advantage. Throughout his life, Friedrich Hayek celebrated the vital role of price in markets. Hayek's deep appreciation for price rests on the fact that as transaction partners negotiate, they have to take into account all the information they have at hand, including their priorities and preferences, and condense them down to a single figure. Let's say a skilled cutler wants to sell a knife that took her a long time to make. She will factor that into the price she wants to get for it. She'll also consider how many knives are available on the market and the price they typically sell for. She'll look at their quality and compare that to the quality of her own knife. Only after she has considered these various elements will she announce a price. A potential buyer will go through a similar process of collecting and analyzing information within the market. Then buyer and seller will either strike a deal—because their prices match—or they'll haggle and negotiate, perhaps gaining further information or changing how they weigh the information they have and adjusting their prices accordingly. If they agree to transact, it sends a signal to the market about the value of the knife. If they don't, that also sends a signal—about the fact that buyer and seller value the knife differently. Rather than spending time communicating a multitude of needs and wants, we communicate a price.

It encapsulates our preferences and priorities into a single unit of information.

The efficiency of the market is reflected in the simplicity of prices as conveyors of information. "In a system where the knowledge of relevant data is dispersed among millions of agents," Hayek said, "prices can act to coordinate the separate actions of different individuals." Price greatly reduces the amount of information that needs to flow through the market; the information is compressed into a single figure for which traditional communications channels are sufficient.

With money, market participants not only know what something is worth on the market. Once we put a value on something, using money, we can trace that value; we can record and compare value over time, thus creating an informational link between the past and the future, and maintaining an external, more objective basis for mutual trust among market participants. Recording monetary values, and thus sustaining trust, is what lets us keep an open tab at our favorite pub and lets dealers maintain a line of credit with their suppliers.

Money may not have been invented to facilitate transactions on the market (scholars of money point to numerous roles money has played outside of an economic context). But it surely has made markets work more efficiently. At first, market currency was a widely agreed-upon placeholder, often a commodity that already carried some intrinsic value. For example, almost everywhere, at some point cowrie shells were used, and in parts of Asia, Africa, and Europe, salt was widely accepted as payment (the term "salary" has *salt* as its root), probably because of its nearly universal demand as a food preservative. The conquering generals of Rome collected grain as taxes. In Latin America, cacao beans were common currency—an early chocolate money with a bit more bite. In North America, animal skins were often used, the origin of the term "buck." Already expressing the value of goods in such standard terms conveyed important information on markets.

But money does not need to be worth something in its own right. Indeed, it is much better when money serves primarily, if not exclusively, as the language in which market transactions are being conducted. When we exchanged commodities such as barley for goods and services, the underlying material could always be kept for its own sake rather than used in a transaction. It had intrinsic value. Gold and silver may not have been directly useful, but these precious metals were rare and shiny; much like diamonds, they turned into sought-after possessions. With the shift to base-metal coins and paper, we moved away from denominating value through a good that was intrinsically valuable. Initially, currency-issuing nations believed they had to prop up their money by guaranteeing that it could be exchanged for gold or silver at a fixed rate. When that practice ended, around the beginning of the twentieth century, money became purely informational. Today, money is moving from the physical to the virtual—the digits signaling a transaction in our bank accounts, the bits denoting an entry into Bitcoin's ledger—further emphasizing money's informational role.

In our daily lives, we may often overlook the informational function of money and price. After all, we are usually far more interested in completing transactions—getting the food to feed our family, purchasing the home to shelter us and our loved ones, or buying the car to get us around—than in focusing on the details of the transaction process. And yet, without the flow of information that money and price enable, we would be ignorant about what others have to offer on the market and incapable of comparing and evaluating quickly and with ease. Money and price are the infrastructure, the conduits of information, that make the market work.

But money and price do much more than streamline information flows; they also simplify transactional decision-making. If we have difficulty comparing and evaluating goods and services across many dimensions, the shift to a price—a simple figure—eases our cognitive load. Just imagine for a moment a world without money: say you want to purchase a loaf of bread, and one baker offers it to you in exchange for a pound of butter, whereas another wants a crate of apples. How would you go about comparing these two offers? In fact, how would you even have gotten the information on what the bakers want in exchange for their goods? If markets have great potential for coordinating human activity but are saddled with the practical problems of costly information flows and cognitive information overload, *money-based* markets realize this potential by reducing flow and simplifying processing of information to an acceptable level.

It is thus no surprise that money-based markets have been enormously successful and form the core of economic activity in most nations around the world. In fact, money-based markets are indelibly woven into the social fabric of nearly every culture on the planet. We're so trained to think in terms of price that when we hear about a new product or service, we almost instinctively ask for its price in order to evaluate and categorize its relevance and value to us. We have become so infatuated with markets and money that we have introduced them in areas quite remote from conventional economic activities. We purchase "winks" to indicate interest in another person on online dating sites. Firms buy and sell pollution certificates to manage fossil-fuel use. And we set up so-called prediction markets to pool (through money and price) available information on everything from Hollywood box-office receipts to the outcomes of presidential elections.

In every one of these marketplaces, price is the key enabler. Consider prediction markets. When participants share their forecasts of a future event, they essentially pool all the information they have. But how do we know what information is accurate and relevant, and what isn't? Averaging all the information we have will not necessarily lead us to the truth. Asking more people and counting views equally is no surefire method to get closer to the truth, either. As the Marquis de Condorcet showed more than two hundred years ago, adding more people only helps when each new person has a better-than-even chance to know the truth. However, when a prediction market allows participants to trade with real money, the overall prediction often improves. That's because those who are confident about the rightness and relevance of the information on which they've made their prediction are more willing to put more money behind their "bet" to maximize the payoff they get if they turn out to be correct. They put their money where their mouth is. As a result, their transactions signal the perceived quality of their information, and their view is weighed more heavily. This does not guarantee that all of the predictions in the market will be correct—far from it—but it is eminently better than giving equal weight to every bit of information.

Google's experiments with prediction markets are but one real-world example of the power of combining markets with money to generate more accurate forecasts of future events. Since 2005, employees at the company have been asked to answer questions about potential developments in the tech industry and the world in general. For instance, they may be asked,

"How many users will Gmail have at the end of the quarter?" or "Will a Russia office open?" and are offered a range of defined responses. Employees participating in the prediction market are given a wallet of "Goobles" to spend on their answers. If they choose the correct answer, they earn profits in Goobles, and at the end of each quarter, everyone can trade in their Goobles for raffle tickets. Thus, market participants have an incentive to spend Goobles only on predictions when they think they have a pretty good sense of knowing whether they'll come to fruition, because that increases their chances for a reward. The price incentive works: the markets have proved quite good at gauging the probability of events related to Google projects, thereby facilitating the flow and processing of information.

These and similar experiments have bolstered the view that the combination of markets and money offers an outstanding way to coordinate human activity. Much time and effort has been spent on improving money-based markets by enabling price information to flow even faster, by making price comparisons even easier, and thus, generally speaking, by lowering the overall cost of the system. When the first issue of the product-comparison magazine *Consumer Reports* was published in 1936, as the world struggled to recover from the economic distress of the Great Depression, its founder believed that there needed to be more aggressive reporting than there had been in the past—more information flow. Newspapers and specialty magazines followed suit, covering everything from the most effective laundry detergent to the best cars, cameras, and computers in various categories and classes. These information intermediaries provided comprehensive reviews and extensive tables breaking down the various product features and components and comparing them side by side. But for all their detail, the reviews almost always appeared under a big, bold headline that stated the case very simply: they listed the top three to five items based on value—the "biggest bang for the buck." Money and price were just too obvious, and readers too well accustomed to their alluring simplicity to push those concepts aside.

Internet price comparison sites and apps that let users find the best deal in absolute or relative terms—including Price-Grabber, Which?, Confused.com (for cars and insurance), Kayak (for travel), and of course Google Shopping—are digital descendants of these information services. So, too, are browser plug-ins and apps such as InvisibleHand and PriceBlink, which can search in the background as you visit Amazon, Walmart, and other retailer sites and notify you if a lower price is available any-where on the Web. They, too, focus on price, taking for granted that the less it costs to discover and compare prices, the lower the overall cost of transacting on the market; and everyone wins.

PRICE-BASED MARKETS ARE THE ESTABLISHED ORTHO-doxy. We are accustomed to them. They do the job. But con-densing countless dimensions of information into a single figure hardly seems the right choice for an *information* age, for an era characterized by astonishing improvements in our ability to communicate and process lots of information.

A system based on money and price solved a problem of too much information and not enough processing power, but in the process of distilling information down to price, many details get lost. Just as a tiny thumbnail JPEG image on the Web offers only a very coarse representation of the original, but is the best we can do given the constraints of technology, we embraced price because we had not found a way to devise another means for decreasing the cost and difficulty of handling richer information flows. But price is compromised by the very fact that it abridges the information available to the market.

For example, your willingness to buy a pair of shoes at a particular price may reflect the urgency of your need, how well you think the shoes are manufactured, and how much you (and your peers) like the style. To some extent it is also a product of how much money you have available to spend at a given moment in time rather than, say, a week or a month in the future. In theory, these nuances are reflected in the amount of money you're willing to pay for the shoes. But no seller can intuit from that figure how much weight you assign to each of those factors. Usually the best a seller can do is analyze total sales by price and volume—that is, in a huge, amorphous aggregate—and adjust prices up and down in an effort to match demand and move inventory.

Say that you spot a pair of shoes in a store window. The style is exactly right for you, but you're not that happy with the color. You'd be willing to buy them as is if the price were just a bit lower. Or if they were available in the color you like, you'd be willing to pay more than the advertised price! Frustrated, you walk away—not knowing that the same style in the color you want can be bought at a shop you've never visited. In another scenario, perhaps the style and color and fit are perfect, but you don't have enough money to buy the shoes on the spot. Two weeks later, flush with cash, you return to the store and discover that the seller had reduced the price because the shoes weren't selling fast enough. Now the shoes are out of stock in your size. If the woman who owns the store had known you well enough, she might have been happy to sell you the shoes on your first visit with the promise that you'd pay her within two weeks, especially because she would have gotten a higher price from you than she got from the eventual buyer. In these cases, the market outcomes are inefficient, because price does not adequately convey enough information about the buyer's and the seller's priorities and preferences.

Over the centuries we have developed ways to deal with some of the consequences caused by the lack of detailed information in a price. For example, if you want to buy something immediately but don't have the cash, you can choose to use credit (if you are creditworthy). Or, with the help of a smartphone, we may search for and discover another shop stocking what we want in the right size and color. Likewise, producers and sellers conduct surveys to determine which aspects of a style—the brand name, the color, the shape, the fit, and so on—are most appealing to their preferred customers and adjust their production levels and prices accordingly. But even though these tools are useful in teasing out the different components of price, they also increase the cost of a market transaction.

Worse, such information reductionism also fails to significantly reduce the difficulties of choosing that afflict every human being participating in the market. Having less information to process does not automatically lead to better decisions. In fact, by condensing information into a single number, we become vulnerable to several well-documented biases that plague our decision-making. Smart marketers exploit this, attempting to distract us from rational evaluation and refocus us on price. Prices ending in nines are good examples, making us believe that something is cheaper than it actually is.

In January 2010, Steve Jobs took to the stage in his familiar black mock turtleneck to announce the iPad. He asked the crowd, "What should we price it at? Well, if you listen to the pundits, we're going to price it at under $1,000, which is code for $999." The price flashed up on the screen behind him. But, Jobs continued, Apple had aggressive cost goals for the iPad, and the company had met them. Accompanied by the sound of shattering glass, the $999 figure was replaced on the screen by $499— the retail price of the first iPad. This wasn't merely a glittering spectacle on Jobs's part; he was "anchoring" the value of the iPad in people's minds to an artificially high price point, prompting them to view it as relatively affordable, regardless of how well its features compared to those of similar products.

We like to think that price allows us to compare apples to apples, but behavioral pricing expert Florian Bauer, of the consulting firm Vocatus, maintains that sellers often use price to deliberately obscure information that would improve market efficiency. This can make us *think* we are comparing apples to apples when we're really asked to compare apples to bananas or oranges. Steve Jobs used this trick when introducing the iPad.

Our reliance on price thus can lead to inefficiencies in the market that hinder our ability to coordinate. When companies devise different bundles of goods, each of which is assigned a single price, for us to compare and choose from, we are so conditioned to focus on price that we give less, if any, weight to the underlying differences among the bundles. Unfortunately, as a result, our decisions are flawed: we agree to an expensive purchase when a cheaper option is available, not despite but *because of* the simplicity of price and the fact that it plays to our biases.

This kind of manipulation has tangible costs. Although the pricing of its new tablet is important for Apple, the flawed choices that ensue are limited in number and effect—but when too many market participants fall victim to the same flawed decision-making, economic disaster may ensue. The subprime mortgage crisis of 2007–2009 has often been characterized as the result of unethical bankers colluding with corrupt analysts at rating agencies to sell risky investment products to ignorant investors, while regulatory agencies looked the other way. There's certainly a lot of evidence for such a view. But there may be another way to interpret this unprecedented obliteration of capital—as a toxic combination of opaque information and deeply flawed human decision-making.

Around the beginning of the new millennium, "innovative" financial institutions began to bundle together subprime mortgages—those carrying a higher risk of default—with other mortgages into securities. The elevated risks associated with the resulting new products weren't exactly a secret, but they were captured in technical language and buried deep in public filings that very few people ever cared to read. Rating agencies, tasked to read the fine print and to evaluate risk, failed to act as "canaries in the coal mine." Without easy access, the available information did not get adequately reflected in the price of the securities. Investors, meanwhile, who were longing for healthy profits in what seemed like a robust housing market, had no obvious reason to worry. When a growing number of homeowners began to default on their payments, it caused a domino effect. In the end, trillions of dollars of value had evaporated, caused at least in part by obstacles in accessing information and blunders in using it. The subprime mortgage crisis is also an indictment of conventional money-based markets and their inability to foster appropriate information flows and facilitate the transformation of this information into well-grounded decisions.

Money eased the exchange and evaluation of market information for many centuries, by collapsing a great deal of it into price. But in large part, those coveted greenbacks paper over the fundamental challenge of taking highly condensed information and translating it into transaction decisions. Money-based markets are fraught with inefficiencies, and these are felt in how well or how badly the market fulfills its promise of coordinating human activities to everyone's best interest. Today, thanks to a number of recent innovations, the market is poised to evolve, leaving behind the straitjacket of money and price, of constrained information flows and crippled decision-making.

$-4-$

DATA-RICH MARKETS

"NOTHING ANYONE DOES WILL SEEM THAT CRAZY anymore." That's how Jason Les described playing poker against Libratus.

Jason Les is one of the world's top players of heads-up no-limit Texas Hold'em, a version of the game with no limits on bet sizes that has made some people rich and impoverished far more. Soft-spoken, like many professional poker players, Les is a guy with strong analytical skills. In January 2017, he and three other poker pros sat down at their high-stakes tables in Pittsburgh's Rivers Casino for a showdown with Libratus, poker's new whiz kid. They played 120,000 hands one-on-one against Libratus over the course of three weeks. Like the very best poker pros, Libratus remained cool despite the pressure: no "flop," no "turn," no "river" rattled the newcomer. But unlike a typical poker celebrity, Libratus wasn't flashy. That's not on the menu for a machine learning system combined with lots of data and housed on a supercomputer at Carnegie Mellon University (CMU).

Libratus's human opponents worked hard to detect quirks and patterns in the computer's style. Two years earlier, in 2015, four poker pros, including Les, had competed successfully against Libratus's predecessor, Claudico. Built by Tuomas Sandholm's team at CMU (which also built Libratus), Claudico had difficulty calculating the likelihood that the pros were bluffing. This led it to place suboptimal bets in a significant number of deals. Each night, while looking through the printout of the hands they played, Les and his fellow competitors spotted weaknesses in Claudico's strategies. The following day, they took advantage of them.

Against Libratus, however, their approach was ineffective. Libratus was just getting better and better as the tournament advanced. As Les noted, "We did have the impression Libratus was adjusting to the way we played over the course of the competition. At the end of the competition, we discovered that it was improving, but not in the way we thought. It was learning all the unusual bet sizes we were trying, plugging its own holes every night as the event progressed."

Libratus had played trillions of hands against itself over several months to prepare for the 2017 tournament. As the system learned, its ability to detect human bluffing had dramatically improved, allowing it to find the optimal bet that would win any given hand—because its opponent either folded or held weaker cards. Reflecting on Libratus's behavior, Les said: "It is completely unaffected by results and always consistent in strategy. If we were to describe a human this way, we would call him a machine." Of course Libratus has no emotions, so it doesn't hesitate to place huge bets, even when it has a bad hand. In 2017, the humans had finally met their match: Libratus racked up more than $1.7 million in chips and won the tournament decisively.

Poker is a beautiful game, not least because it involves a combination of psychology, probability, and game theory. Excellent recall, numeracy, and rational thinking form the foundation of every good poker player's skill set. But these attributes aren't enough. Professional players must also possess superb communication skills. At the table, they must read the "tells" of their opponents—not just how they sit, squint, or hold their cards, but also what their betting behavior signals (the latter is what Libratus and its opponents had to focus on). At the same time, they must offer very few tells themselves—and sometimes fake them in the hope that their opponents might take the bait.

In these and many more ways, poker is much closer than more symbolic, abstract games such as chess and Go to our real-world experiences of strategizing, signaling, negotiating, and transacting on the market. The elements—wagers of real money, the buzz of strategy and bluff, the subtle dynamic of reading and sending signals—feel very familiar. Thus a computer that beats champion poker players at their own game stuns us. How unique are humans in their ability to wheel and deal, to strategize and communicate?

Libratus's impressive victory indicates that a computer may be capable of transacting in the marketplace better than we can or, at the very least, that a computer can greatly aid humans in conducting market transactions—not because it runs more calculations per second than our brains do but because, unlike humans, its decisions remain unclouded by human cognitive constraints.

Consider betting strategies: most human poker players do not see the logic in betting a huge amount to capture a small pot.

If a player places a huge bet for a small pot, the opponents generally assume the bettor doesn't understand the game or is bluffing badly, because the bet will almost certainly stop the opponents from betting more and will limit how much money the bettor can win. Libratus, however, regularly placed huge bets, and in the end the strategy paid off richly. Numerous cognitive biases—from misjudging risks and sticking to a strategy even in light of new information to an imprudent disregard for small wins—cloud a human's decision-making and militate against huge bets for small pots (among other things). In contrast, Libratus reevaluated its strategy after every move, spent nights methodically revisiting the hands of the day, deducing behavioral patterns in its human opponents, and honing its own strategy to exploit them. And Libratus stubbornly crunched through huge volumes of data that overwhelm human decision makers. In consequence, Libratus won far more often than any of its human opponents, even if the average individual wins weren't spectacular.

By combining a reevaluation of strategy with learning from the past, Libratus could "see" the tournament not simply as a large number of individual encounters, but as sequences of games that reveal an opponent's behavior and weaknesses without locking Libratus into a prohibitively fixed model of human behavior. It's a kind of strategy that smart negotiators use across many rounds of negotiations, and one that shrewd merchants employ, especially for repeat market transactions. Unsurprisingly, Professor Sandholm, Libratus's designer, envisions a commercial version of the system to bargain on behalf of consumers and businesses in complex market transactions. But that is only the beginning. The triumph of Libratus foreshadows an even more fundamental shift in our economy, and, just as with Libratus, the driver of that shift is data.

As we have shown, markets are amazing social innovations that enable us to coordinate our activities with each other efficiently—in principle. In practice, they suffer from limited information flows. We rely on money and price to reduce the amount of information that needs to be communicated and processed. But information condensation means that market participants aren't always able to share their preferences comprehensively or to weigh them appropriately in their decision-making. Price may solve the problem of too much information, but it causes us to choose badly. Our fixation on price has hampered the market's ability to do what it does well: coordinate.

The answer to this problem isn't digital payment, or virtual money. That might speed up existing information flows, or make them cheaper, but information would still be compressed into price, eliminating valuable detail. The solution is not to fiddle with money but to replace—or at the very least complement—its informational role with rich and comprehensive streams of data. Data is the new grease for the wheels of the market. It helps market participants to find better matches.

Thus, the most immediate and obvious difference between conventional markets and data-rich ones is the volume and variety of data that flows among market participants. Rather than being restricted to the information trickle around price, in data-rich markets participants would aim to convey and act upon the full gamut of preference information, utilizing the market's informational structures to communicate all this data at low cost.

In theory, we could have utilized more and richer data in analog days. But it would have been very costly. Thanks to digital networks, massive amounts of data now can flow quickly, easily, and cheaply between transaction partners, whether they are near each other or thousands of miles apart. But just widening the data pipes alone, as much as that might overcome the "dearth-of-information" challenge, would likely lead to an information overload for market participants. How would we, so accustomed to and focused on price, compare products across many dimensions and then identify the right match? How would we express our multiple preferences swiftly and easily?

Money and price may be an information straitjacket, but escaping it requires not just very different ways of communicating information; it also necessitates a step-change in how we translate information into decisions. We not only need vastly more data, but also the right methods and tools to work with that data. It is precisely the absence of such methods that has kept money-based markets in place in the early decades of the digital age. Things are changing, however. A recent confluence of advances in data-handling is finally enabling us to leave behind the limitations of money and price and embrace data-richness on markets.

Three key technologies are crucial to this reconfiguration of markets. They allow us to (1) use a standard language when comparing our preferences, (2) better match preferences along multiple dimensions so that we can select the optimal transaction partners, and (3) devise an effective way to comprehensively capture our preferences. All three technologies have in common that they facilitate the translation of rich data into effective transaction decisions. Underscoring the central role of data, these technologies not only improve our ability to choose based on data, but the technologies themselves are founded on data. Together, they provide the foundation for an economic revolution.

When baby boomers went on vacation, they had to thumb through inch-thick hotel brochures and meet with travel agents to confirm whether the brochures' slick marketing copy and glitzy photos were accurate. If they were fortunate enough to know somebody who had stayed at a particular hotel before, they could rely on that person's recommendation. But that was the exception rather than the rule. Today, by contrast, we choose our accommodations after sifting through a sea of information—customer ratings, journalists' reviews, photographs posted online by previous guests—and we can quickly compare hotels by location, amenities, and quality of service. We can even take a virtual road trip to the place, thanks to Google Street View. And when it comes to price, online comparison will easily tell us when and where to get the best deal.

Likewise long gone are the days when we rented a car or looked for a ride-share on the basis of price alone. BlaBlaCar boasts more than 40 million members in more than twenty countries and allows riders and drivers to get matched along multiple dimensions, including their self-reported level of chattiness—from Bla ("watches the scenery go by") to Bla-BlaBla ("won't keep quiet"). Hence riders are more likely to take other information into account when selecting a ride. The approach has people chatting: at the time of this writing, 4 million people book rides through the company every month.

This information convenience is pleasing for its ease of use and accessibility (at least most of the time). Our travel transactions are more efficient because buyers and sellers can match their preferences more precisely. Of course, such richness of data isn't only springing up in the travel industry. When we shop online for anything from books to electronics to clothes, we have at our disposal scores of characteristics to consider as well as sophisticated searching and filtering tools that enable us to browse, research, and compare products.

What makes this work isn't the speed, low cost, or storage capacity of the technology we use. It's not even simply the increased volume of available information. What fuels better matches is that we have an efficient way to label and categorize information.

Let's say you're shopping for a new shirt. You go online to the site of your favorite retailer. You click on "Shirts," and the site gives you hundreds of choices. But you can filter these choices—or filter out the ones you don't want—by selecting your preferences among a staggering number of factors: size, fabric, color, fit, sleeve length, type of collar, and perhaps even brand. So if you want a boatneck cotton knit top with three-quarter-length sleeves in size eight in either blue or turquoise—preferably one that's on sale—there it is. And if it isn't there, you can move on to another source. How can an online retailer provide you with that much information about its shirts? By labeling each product with data that describes each garment's characteristics. This requires, however, that all products of a particular kind, say "shirts," are labeled using the same set of categories. These categories are data, too; but they are data about data, or *meta*data.

This isn't new. Since the time Assyrian clay tablets were first affixed with labels describing their content, information about information has been important. Today, efficient labeling is essential. Without it, we have little hope of finding anything online. By the same token, the process has gotten harder. In the old days of relational databases, data was neat and tidy, because every data field was clearly defined, down to specifying the exact format of the field's content. Since the late 1990s, however, this orderliness has been challenged by the exponential growth in digital information, much of which does not fit neatly into a database field: it comes in the form of e-mails, Web pages, images, and audio and video files.

Consider the case of YouTube, a market for video content in which uploaders (i.e., sellers) transact with viewers (i.e., buyers), often financed by a third group of market participants, advertisers. To assure that videos will be watched, viewers need to be able to find content easily; for the same reason, content providers need to be able to make their content quickly discoverable. The title of a video and the date and time of its upload only go so far. Adding labels and keywords to the video is only as effective as the uploader's ability to select the right keywords.

Commercial content providers face the same problem. A sports network such as ESPN broadcasts and records hundreds of thousands of hours of video footage every week. Although some fans may want to watch an archived sports event from start to finish, many will want to go straight to the most important moments—replaying LeBron James's decisive chase-down block in Game 7 of the Cavaliers' comeback in the NBA championship in 2016, for example, or Dave Roberts's ninth-inning base steal in Game 4 of the 2004 American League Championship Series, which put him in position to end the "curse of the Bambino." To ensure that these moments are easily discoverable, ESPN has been relying on human labor, employing dozens of people to watch multiple sports events simultaneously in real time and to manually tag every play and interaction.

If ESPN were letting staff tag the videos in any way they wanted, the project would not be all that different from the hit-or-miss labeling on YouTube—just an improvement in scope and scale. But these taggers have also been trained to use a well-developed hierarchy of keywords, what experts in the field call an "ontology," as they label the videos they're watching.

Sports lends itself to ontological systems. Every sport—from archery to wrestling—has defined sets of rules, not only for the players but also for the competition itself. The same is true of books, electronics, and appliances. Whenever there is a clearly delimited set of parameters, it's easier to discover the products most appropriate for any given consumer. Because publishers have more than a century of experience classifying books into discrete categories, following the Dewey Decimal or Library of Congress systems, if you'd like to buy a book on the history of women during the Civil War, you can probably find it. Indeed, one reason Jeff Bezos started Amazon as an online bookstore in 1994 was because publishers' seasonal catalogs had recently been digitized, and he planned to build his company from the foundation of that data.

The same foundation allows Amazon shoppers to select, filter for, and compare consumer goods not only according to brand, price, and buyer reviews but also according to many other less obvious characteristics. For washing machines, for instance, there is information about how a washer opens, its color, its size, and, in some European markets, its load capacity and energy efficiency. Similar information dimensions exist for numerous other products, such as TVs, hard drives, and microwave ovens. Labeling the features of electronics is often relatively straightforward: the manufacturers either provide sufficiently rich data to the online retailer, or the online retailer adds the data itself as the ontology is fairly obvious. Generally speaking, there are more markets with rich information flows for product segments that lend themselves to simple and accepted ontologies.

By contrast, developing an ontology for a general marketplace is much more difficult. That's why finding YouTube videos is far more hit-or-miss than shopping for washing machines at Amazon. How do you search for a concept—say, a video on how to do somersaults? YouTube cannot yet match the depth and breadth of the keywords that are standard at ESPN, simply because humans have not yet been able to come up with an easy-to-grasp general-purpose ontology that everyone can understand quickly and apply flawlessly.

EBay has long been struggling to provide a comparable level of discoverability in its marketplace. Unlike customers who use Amazon's conveniently rich filters, buyers on eBay often have had to search for words in product titles and descriptions, then scroll through page after page of results. This is the legacy of eBay's start as a marketplace where anyone could sell anything, including goods that were in many ways unique, whereas Amazon began as a seller of products (books) in a single category with a well-developed product ontology. Over time, the lack of an ontology in a market reduces the number of transactions that take place, because people have trouble finding a match even when one exists. Without clearly usable filters to ease discoverability, a market's efficiency plummets.

Because success in many marketplaces hinges on enabling a rich flow of data, there is considerable economic pressure to develop efficient labeling strategies. Madi Solomon, an expert in such data, emphasizes that the key lies in finding the *right* ontology. She knows how difficult this can be—she describes herself as "coming from the salt mines" of data, having worked as the corporate nomenclature taxonomist for the Walt Disney Company (which owns 80 percent of ESPN) and then as the director of data architecture and semantic platforms for the educational publisher Pearson. In the future, however, Solomon thinks identifying the right ontology will require less human ingenuity than hardheaded data analysis: data will drive data ontologies.

Considering how much depends on getting labels and categories right, as well as how relatively limited our capabilities are so far, it's easy to see why data ontology is a hot field for information technology start-ups and an important tool for transforming money-based markets into data-rich ones. The massive data project underway at eBay aims to improve cataloging for products on offer, increasing the rate of discoverability from 42 to about 90 percent. They are already acquiring and working with a number of data ontology start-ups, such as Alation, Corrigon, and Expertmaker, to automatically categorize product information. Other marketplaces are following suit, racing to put the data infrastructure in place that will enable a rich, multidimensional flow of information. Without it, markets, offline and online alike, will remain locked into the conventional focus on price.

We are already enjoying data-rich markets in numerous sectors, such as travel, ride-sharing, and electronics. But the richer the information, the more difficult it is to process it—to weigh each dimension based on our preferences and select the optimal transaction partner. Translating an avalanche of information into decisions is hard. Who hasn't gotten overwhelmed by too many filters and options when searching for airline flights on online platforms, such as Expedia, or for a place to stay on Airbnb? Even if all offers are plainly visible to us, identifying the best one is often difficult. The challenge is information overload, including having too many options to filter and select, and thus to identify the optimal match. Fortunately, here, too, technology can help.

In conventional markets focused mostly on price, matching preferences of a buyer and a seller is relatively trivial. All preferences are condensed into the price a buyer is willing to pay and a seller is willing to accept. Bringing the two together is supposed to happen pretty much by itself, as long as buyers and sellers state their various preferences in terms of price and as long as there are sufficient (and sufficiently diverse) market participants. In practice, valuable preference information gets lost, perhaps because market participants fail to reflect correctly all their preferences in price, but also because others erroneously deduce preferences from a price. Under those conditions, something that looks like a good match, in fact, is not. We may think the market works, but in truth it doesn't, and it leaves everyone worse off.

Data-rich markets have the advantage of not deducing preferences from price. They offer another advantage over price, as well: not only do individuals have multiple preferences regarding a potential transaction, they also likely weigh different preferences differently. When preferences are condensed into price, two preferences weighed equally may yield the same price point as two preferences weighed very unequally (one very high and one very low, for instance). In data-rich markets, the raw preference data, including relative weights, is available, but it requires a matching process that is smart enough to take these multiple dimensions of preferences and their relative weight into account. Doing this manually is challenging for most humans, and it requires time and effort that few may be willing to invest. Data-richness would all be for naught if the detail in the preference data isn't acted upon and used to identify the best match.

Fortunately, over the past few decades mathematicians and economists have been hard at work developing algorithms to evaluate sets of multiple preferences and their relative weights and to identify best matches. Although the actual process is quite technical, at its core it isn't too dissimilar from analyzing and matching patterns in data. It is the same technology we use to manage our photo collections to find pictures with certain features, or to have our smartphones "understand" voice commands, or to make the health apps on our smart watches detect telltale signs of a dangerous heart condition. Because preference data is just a data stream forming a particular pattern, we can adapt pattern-matching algorithms to help us identify optimal transaction partners. This isn't simple by any measure (choosing exactly what to compare against what isn't trivial), but thanks to better algorithms, improved in large part through huge amounts of training data, the task has been getting easier. In data-rich markets, these algorithms are the method by which transaction partners may find each other.

This is a huge improvement over transaction decisions based on price; it enables buyers and sellers to take full advantage of the comprehensive data flows available and helps them translate data into transactions effectively and efficiently. Because of the decentralized nature of the market, information exchanged between market participants is dyadic: after a potential buyer has communicated with a potential seller and exchanged preferences, both know about the other, but not about the entire market. Moreover, market participants may not want to reveal all their preferences to the market. This and similar behavior leads to the information asymmetries we mentioned earlier. Data-rich markets do not eliminate such asymmetries; but because more preference information on data-rich markets generally leads to better matches, there is less of an incentive to keep information from others: vastly improved matching aims to identify the transaction partner that gets the most value out of a transaction, the partner who is thus willing to pay the highest price, arguably outweighing some of the advantages in negotiations that many information asymmetries may offer. In data-rich markets, each exchange between potential partners reveals more information, even if not leading to a transaction, and thus betters the outcome. And advanced matching even helps where information asymmetries persist by carefully orchestrating the matching process to improve overall welfare. Of course, the process is iterative; even if bits flow fast and cheaply, it still takes effort, and because nobody will know every preference of everyone else, transaction decisions, though much improved, won't be perfect.

Some market participants may agree on transactions that further both their interests but leave others worse off. In some instances, the outcome, although individually positive, may be "welfare reducing"—the economists' shorthand for destroying rather than creating overall value. Of course, the cost of not always achieving maximum overall welfare is a small price to pay in return for the vast relative improvement that we get through the individual matching processes, thanks to the shift to data-richness. However, for some very specific types of transactions, especially those that have huge consequences beyond the immediate transaction partners (economists call this "externalities"), we may want to apply lessons from existing markets that must function without price. They work through clever market design combined with a different type of matching algorithm. Think, for example, of choosing which patient gets a donor kidney. Donor kidneys aren't sold (at least legally, although some economists have suggested they should be), so preferences can't be condensed and simplified into a stated price. In such markets, a central clearinghouse often collects preference information from all market participants and uses advanced matching algorithms to connect suitable market participants to transact. The goal is to produce as many suitable matches as possible. This sort of matching, too, has recently improved significantly, thanks to enhanced algorithms and a better understanding of which matching algorithm works best for which type of market. In 2012, two of the world's leading experts in matching, Lloyd Shapley and Alvin Roth, were awarded the Nobel Prize in economics for their theories on the subject.

For transactions with huge externalities, data-rich markets could utilize a similar approach; and the richness of their data streams would facilitate the sophisticated matching that needs to be done by the clearinghouse. But it would require that everyone on the market agree beforehand on a set of principles concerning how the matching will work, and that these principles are strictly adhered to (lest the market participants lose trust in the matching system). Hence, such an approach with a centralized matching authority (although participants retain the ultimate decision whether to join the market or not) is suitable only for highly specific contexts, and in the vast majority of markets, we'll use a data-rich and algorithm-enhanced, but iterative and decentral matching process.

A more pattern-oriented matching based on rich data is popping up in a wide variety of different contexts, and in different forms. Music platforms such as Spotify and Apple Music aim to match listener preferences with individual songs. The same is true of Netflix and Amazon product recommendations. But this is only the beginning. Not all these well-known algorithms employ all the dimensions of preferences available to them. This opens up exciting opportunities for innovative start-ups. Many of them are vying to be the one to offer the next big breakthrough in matching. For example, the London-based start-up

Saberr suggests that personality-based algorithms can help build highly effective work teams. Saberr's cofounder, Alistair Shepherd, uses results from personality surveys to create an algorithm to discover compatibility within a group of people. He tested his algorithm at a competition for entrepreneurs at which individuals not knowing each other were grouped into teams and then surveyed. Shepherd's survey didn't ask anything about the participants' work experience or education. At its first demonstration, the algorithm predicted which team would win the competition as well as exactly where the other eight teams would finish that day. Shepherd has replicated those results by predicting the winners of the eight-month-long Microsoft Imagine Cup as well as the investment choices made by venture-capital fund Seedcamp. Deloitte, luxury goods conglomerate LVMH, and Unilever are among Saberr's clients.

Because better matches benefit not just market participants but also the market as a whole, we are tempted to think of preference-matching algorithms as a service improvement offered by the market. That's what Apple and Amazon, eBay and Alibaba, Netflix and Spotify are aiming for. As marketplaces compete for participants, it's easy to see how better algorithms can translate into a competitive advantage for the market provider. The more markets move away from a focus on price to data-rich matching, the more the race for better matching will intensify. Thus, we can expect matching services to turn into key differentiators on marketplaces. In the long run, however, these competitive advantages will likely diminish as most marketplaces adopt comparable smart matching technology. At that time, matching will have turned into a basic service, a utility that markets are assumed to provide.

By the same token, matching services do not necessarily have to be provided only by marketplaces. One could imagine opportunities for new intermediaries promising better matches to those market participants that share their preferences and related information with them—think of them as partial clearinghouses. If this happens, value creation in the matching process on the market shifts from the provider of the market to the supplier of optimal matches; as a result, the marketplace may turn into a commodified service with most value (and thus most profits) captured by the intermediaries. And markets may discover that they aren't just competing with each other but also with a new group of disrupters focused on matching. We see this unfolding already in financial services, where new data intermediaries, such as PeepTrade, offer more comprehensive information and better matching services than existing trading platforms. They will be able to charge a premium for access to their insights, while conventional market platforms see their services, such as facilitating buying and selling securities, turn into low-price commodities.

But there is yet another element that's needed for data-rich markets to work. Rich-data streams and improved matching abilities are like a car without an engine if they're not paired with a robust, rational way to help market participants express their preferences (and turn them into data).

With data-richness, market participants may learn the preferences of others and pair them using matching algorithms, but how do market participants express their preferences and their relative weight and communicate them to each other? It's a difficult challenge, and solving it is crucial. Nobody wants to transact on markets that require hours of time spent answering questionnaires. Fortunately, here, too, recent technical advances have gotten us much closer to viable solutions. Consider again Amazon's product-recommendation engine: at first glance, it's a matching system. It quite successfully matches our preferences with available products and makes recommendations about what we should order. But that is only half of the story. Amazon captures our preferences not from us directly but from the comprehensive data stream it gathers about our every interaction with its website—what products we look at, when and for how long we look at them, which reviews we read. Amazon looks for unique patterns in the data that reveal our preferences. Identifying such patterns enables Amazon to statistically deduce our wants and needs without having to ask us directly. It doesn't know them exactly, of course, only approximately (and sometimes will make erroneous recommendations); and it does not know why we prefer one thing over another; it just takes into account the fact that we do. But that is sufficient for Amazon to feed its preference-matching algorithm and search for the products we are most likely to purchase.

Amazon's strategy isn't unique; it's representative of Big Data, an approach to data analysis that aims to capture data comprehensively about a particular phenomenon, looking for complex patterns embedded in the data. By concentrating on pattern analysis, it differs from conventional statistics that have been focused on condensing data to its essence, from calculating averages to running regressions. A feature of many Big Data approaches is that the pattern one is looking for isn't defined from the outset; rather, it emerges as huge amounts of training data are analyzed. In the context of Amazon's recommendation system, for instance, this means that the system did not know which data pattern would suggest a particular customer preference; it was only by going through years of past interactions and purchases that the system would discover the most likely one. Because the system learns as it sifts through training data, it is often characterized as an "artificial intelligence" approach, even though that term originally referred mostly to systems that had been fed general rules rather than having them learn through training data. These systems don't understand the data in any human sense; they only identify the patterns they are "seeing," much like Libratus does when it beats the pros at heads-up no-limit Texas Hold'em.

For such a machine-learning approach to work well, two conditions must be met. First, huge volumes of data are needed initially for machine learning systems to train themselves and make explicit what is embedded in the data. For example, Google utilized all text from the Web to uncover the probability patterns of word usage for its language-translation tool. And, second, the system must receive frequent feedback so that over time it can self-adjust based on the specific and changing circumstances, going beyond its initial training. Newer machine learning systems are looking for more than patterns in the data: they utilize feedback data in a more nuanced, differentiated way, devaluing older data for instance, a bit like human memory does.

Feedback is central to any such system, especially when the system is used to assist in critical decision-making. Tesla's CEO, Elon Musk, boasted on Twitter in late 2016 that his company's cars logged many hundreds of millions of miles using Autopilot, Tesla's semiautonomous driving system. Likely, it wasn't simple numbers-bragging that drove Musk to tweet. Autopilot generates and accumulates valuable feedback data that gets sent to Tesla and is used to "train" the next software release of the Autopilot system. Teslas literally get better with every mile somebody drives them.

The same feedback process that keeps Teslas on the road can be used to learn about changing preferences of market participants. If a customer repeatedly buys a particular toner cartridge for a printer from the highest-quality provider regardless of price, that buyer reveals a preference for quality; the system does not need to know why the buyer is relatively price insensitive. When that customer starts buying the cheapest toner instead, it signals that preferences may have changed, and the system will adjust.

Several of today's most powerful adaptive machine learning systems are trained with huge amounts of data initially, then learn to adjust to a specific individual. For example, the intelligent assistants built into some of our devices, such as Amazon's Alexa and Apple's Siri, can convert speech to text because the system has been trained through analysis of billions of audio data points covering a wide variety of pronunciations. Once you start using the assistant, it uses feedback to adjust itself based on your language use and preferences. Start-ups around the world, too, are focusing on teasing out preferences from feedback data through machine learning. For example, Infi, a preference assistant developed in Israel analyzes a wider variety of smartphone and social media data.

In the market, the combination of massive, data-based training followed by adaptive feedback and personalized learning offers the potential for significant efficiency gains. Adaptive machine learning systems can reduce the influence of our cognitive biases in decision-making while still allowing us to be ourselves. Because such systems rest on lots of initial training data, that data represents feedback signals of a very wide variety of individuals. Although every individual is saddled with a unique mix of biases, signals from a large group of diverse individuals may diminish more extreme forms of bias. The cognitive limitations implicit in our preferences will not disappear, but the system may help us revert to the mean—if we want that.

As feedback mechanisms evolve, it will become possible for an adaptive system to identify preference data from less biased sources and weigh that data more heavily. After all, unlike humans, these systems aren't limited in how much they can learn. This could lead to systems that come already preloaded with a robust, comprehensive set of preferences—a smart, even-keeled decision agent that can step in whenever you do not trust your own judgment. But it could also adapt to our specific preferences (and thus also our biases) by observing how we decide and how we react to suggestions the system proposes to us. Over time, the system may become more like us. How much or how little is something that the system, too, might be able to discern implicitly by analyzing our feedback on its decision suggestions: if we liked when it tried to avert one of our biases, it'll note that—and vice versa. In short, such systems have the potential to offer the best of both worlds: to expose us to decision expertise from thousands or millions of other market participants while learning and following our very own preferences and priorities over time.

Such machine learning systems can be useful in any context, not just the market. But some social structures are better able than others to generate the feedback data that's required. Thanks to their decentralized nature, markets produce a unique flood of signals that a machine learning system can absorb and learn from. Each signal—from the exchange of money when a transaction is agreed upon to a minuscule gesture of interest (or lack thereof) as a person scans available options—has informational value. Even the sequence of interactions matters—the order in which the signals occur. The signals are small enough to create a large supply of data, yet the data points are sufficiently connected for preference analysis.

All the necessary elements for the re-creation of the market are in place. Improvements in data ontology help us extract valuable data from huge streams of it and categorize it in many dimensions. Advances in matching algorithms enable us to find and select the optimal transaction partner in the market of our choice. And machine learning systems identify our preferences as they observe us so that we don't have to spend time making these preferences (and their relative weighing) explicit. As our trusted assistants, they advise us as we choose and alert us (if we want them to) when we are making a biased decision. They may even end up making a lot of decisions for us.

When combined, these techniques will make us formidable buyers and sellers, not because we will win every negotiation but because we'll act efficiently, relentlessly optimizing outcomes based on our preferences. They will not only benefit participants but will greatly improve the market as a whole and make it the most efficient place to coordinate human activity.

Each of the technological advances we outlined in this chapter plays a distinct role in overcoming the market's two fundamental challenges: obtaining access to a rich, multidimensional flow of information at low cost, and translating that information into decisions. Data ontologies help the flow of information; adaptive machine learning systems and preference-matching algorithms help us process information. They also reinforce each other in subtle but important ways. Machine learning systems can be utilized not only to tease our preferences out of data; they can also be used to improve preference-matching algorithms and to discover word patterns that will lead to superior data ontologies. Similarly, data ontologies may help us find a better way of ordering our preferences. And preference-matching algorithms can not only assist us in finding the optimal transaction partners, they can also help us identify the most appropriate set of external preferences against which to benchmark our own.

We foresee a time when one marketplace after another will reinvent itself using the advances in technology and the concepts we have outlined. The change is already under way. But it won't be a simple, swift, or linear transition. As marketplaces are innovating, they will have to experiment to discover the right combination of technology and market design that suits the needs of their participants. But once a money-based marketplace has turned itself into a data-rich one, a marketplace built on multidimensional information streams, enhanced by preference-matching algorithms and machine learning, there will be no turning back. We can already see such a development unfold in a market that serves as a guinea pig for reinvention—the market for love.

FOR THOUSANDS OF YEARS, PEOPLE SEARCHING FOR LOVE have asked intermediaries for help. Matchmaking is a very old profession, and many social events exist in every culture that are at least in part designed to give people looking for a soul mate an opportunity to find one. But that opportunity has long been constrained by geographical barriers and lack of information. Our perfect soul mate may live two villages away, but because we don't know that, we will never meet.

This is why dating websites caused an almost immediate sensation in the early days of the Internet. The most popular among them seemed to offer a thick market for love—that is, a market in which a large number of diverse participants increases the chances for success. Eli Finkel, an expert on online dating and professor of psychology and management at Northwestern University, calls the members of this first generation of dating platforms the "supermarkets of love." They assured that the marketplace was teeming with potential partners. Users liked that, but they quickly got overwhelmed by the effort necessary to find a mate among the masses—to identify "their" needle in the haystack.

In response, dating sites developed elaborate surveys and questionnaires to tease out preferences and help customers identify the best possible matches quickly and easily. Essentially, the sites switched to multidimensional information streams and implemented preference-matching algorithms. It seemed a sensible move, but it failed badly. It took participants hours to answer hundreds of questions about themselves, and the resulting matches were only marginally better than a random walk in the supermarket of love.

Online dating sites reacted as conventional markets might have done. New competitors assumed the problem was cognitive overload and decided that rather than more information, people wanted less. Much as traditional money-based markets have aimed to reduce the complexity of preference matching by headlining price (to give users a sense of comparability), more recent online dating platforms, like Tinder, narrowed the necessary interactions down to a single dimension—desirability. Swiping left and right is its purest form. By condensing the decision to a single dimension, the process of matching gets easier. But just as comparing prices doesn't tell you everything you need to know when making a transaction, reduction to a single dimension does not guarantee a successful outcome when dating.

The problem is that in the attempt to improve mediocre results, dating sites dumbed down their service. Shifting to multidimensional information, it turns out, was the right move, but it wasn't sufficient by itself. As Professor Finkel explains, the questionnaire process was too simplistic, looking just for similarities (or opposites), and it used the wrong data to boot. What's necessary, in our parlance, are improved preference-matching algorithms and a better data ontology that capture how people *relate* to each other. Rather than asking customers to spend hours answering questions, future dating services will use machine learning systems that deduce the necessary relational data from video, photos, speech, and perhaps even wearable tracking devices. They will register when we smile or blush as we interact with somebody we like and know when our hearts begin to beat in sync. If a system gathers our preferences without much effort on the user's part, and combines them with the right multidimensional flow of information and the appropriate matching algorithms, that will lead to a substantial increase in successful matches. These next-generation dating markets will be far more sophisticated and efficient—although perhaps less exciting. The necessary techniques for reinventing online dating markets already exist; they just need to be combined correctly.

What's happening in online dating is happening in other markets as well. Some markets are ahead; others still lag. But it is a change no marketplace that wants to stay in business will be able to resist. In just a few years, we'll have at our disposal powerful data-rich systems that know us well enough to offer meaningful assistance with our market transactions. In return, having spent less resources and less time to get a better match, we'll reap a handy efficiency dividend. But that's not all.

For the first time in human history, we'll have a choice about whether and to what extent we involve ourselves in certain decisions that are the bread and butter of human coordination, but not exactly the recipe for a satisfying life. We will be able to direct a machine learning system to do the boring stuff and reserve those decisions that give us the most joy and pleasure for ourselves. We will deliberately give up some of our choices so we can focus on the choices that matter most to us. As a result, we will be able to disentangle the need to decide from the pleasure of choosing.

But as we remake the market, infusing it with rich data, we also have to understand, and then rethink, the role of the firm.

— 5 —

COMPANIES AND CONTROL

Ever since it was founded in 1994, Amazon has been in the disruption business. Today it is hailed as a one-stop, one-click bazaar that infers your product preferences by analyzing your digital shopping carts, racking up annual revenues of more than $100 billion. First bookstores, then pet stores, then shoe stores—retailers in one sector after another have struggled against the vast online marketplace. It didn't simply offer a larger catalog of goods than any brick-and-mortar store could stock but also invited individuals to sign up as sellers, from artisanal cheese makers to self-published authors. But opening its virtual doors to individuals wasn't the only way in which Amazon embraced the model of the market. It also allowed buyers to have an active voice, by rating and reviewing products so that future buyers would have easy and direct access to relevant market information when deciding what to purchase. If there ever was a poster child for a successful digital organization that behaves more like a market than a firm, it seems the Everything Store would have to be it.

Yet this is far from the full story. In a number of important ways, Amazon embodies the hierarchically organized, command-and-control structure of the firm that has coordinated much of the world's phenomenal economic growth over the past few hundred years. Amazon is structured as a traditional firm, and Jeff Bezos is a traditional CEO, looking for ever more efficient and effective ways to control every aspect of his organization, and comprehensive information is his tool of choice.

Bezos's attention to every pixel is legendary. In 2011, a former Amazon engineer named Steve Yegge garnered international attention when he shared his thoughts on his ex-boss in a rant on Google Plus that he had not meant to post publicly. "Bezos is super smart; don't get me wrong," Yegge noted. "He just makes ordinary control freaks look like stoned hippies." On Glassdoor, the website where employees can anonymously rate their employers and managers, Amazon has ranked notoriously low in job satisfaction compared to other darlings of Silicon Valley. Many reviews complain about the demands placed on employees and declare that they have no autonomy. A 2015 *New York Times* investigation of working conditions among office staff found that employees are "held accountable for a staggering array of metrics" about different aspects of the firm's operations—running about fifty pages long—and are asked to explain detected inefficiencies in weekly and monthly business review sessions.

Processing that much information puts a drain on morale. "If you are a good Amazonian, you become an Amabot," one employee told the *Times* reporters. The best Amabots claim to put in one-hundred-hour workweeks in order to pore over the data and answer any and all questions posed to them about how the information should influence decisions. Others grow frustrated, burn out, and move on—or, if they are among the bottom 10 percent of the company's performers, get a warning or get fired. The metrics allow Bezos to control staffing up and down the hierarchy without observing performance firsthand.

When the *Times* exposé appeared, it struck a nerve, garnering 5,858 comments online, the most in the website's history up to that point. As the *Economist* noted, many of the commenters "claimed that their employers had adopted similar policies. Far from being an outlier, it would seem that Amazon is the embodiment of a new trend"—what the magazine branded "digital Taylorism," after the scientific management principles of Frederick Winslow Taylor. It seemed that new technologies were ushering in a supercharged version of command-and-control, fueled by data about employees, processes, products, services, and customers. But why would a celebrated marketplace innovator like Jeff Bezos embrace the centralized structures, rules, and behaviors of the firm to manage the vast majority of his business empire rather than developing technology to capture the decentralized magic of the market? Are Amazon and other digital unicorns (as well as many of their smaller brethren) not realizing that the data-rich marketplaces they are helping to evolve also have consequences for them as firms and may force them to rethink their own raison d'être, or at the very least their organizational setup?

To answer that question we need to understand how information flows in firms and is translated into decisions; and how, over time, new tools and methods have repeatedly helped the structure of the firm to evolve.

THE FIRM CAN BE MANY THINGS, INCLUDING A LEGAL entity to raise capital, bundle risks, and help disentangle management from ownership. For our context, though, as we have explained, the firm, much like the market, is a mechanism to enable human coordination. Much like the market, the firm aims to offer such coordination at low cost. And much like the market, firms are designed, at least in principle, to scale well—to continue to coordinate efficiently as they grow (or shrink). The key difference between the market and the firm is how decisions are made and by whom. In the market, decision-making is decentralized and spread across all market participants. In the firm, decision-making is far more centralized and vested in a relatively small number of individuals. This difference is intertwined with and reflected in how each one of them handles the flow of information and its translation into decisions. Understanding how firms might be faring as data flows and data technologies change thus requires us to examine how this is happening currently. It is an astounding case of repeated innovation—technical, but also organizational and social.

Only with sufficient, timely, and accurate information from all parts of the organization do a firm's leaders have the necessary raw material for making decisions. For centuries, this insight has driven the development of effective methods for reporting in general, and the invention of accounting in particular. Often glossed over by the public, these methods are among the most crucial ingredients in the rise and success of the firm—at least as important as mass customization and globalization. Initially, they provided data about a firm's financial performance, but over time they got broadened to comprise all aspects of a firm's activities.

Having a more comprehensive view of what's going on is the necessary foundation for shifting the aim of reporting from holding an organization accountable for the past to utilizing the information as the basis for strategic planning into the future. Data replaces gut feelings; the aspiration toward rational management supplants the reliance on idiosyncratic decision-making. And though the focus here is on the flow of information, the ultimate aim, of course, is better decisions.

In the early days of the firm, internal reports on business activities were narratives—stories about what happened in a transaction, told in person to one's partners, or, increasingly, conveyed through written diaries to bridge space and time. Few of these "accounts" included much in the way of numerical calculation, because most early numbering systems were too complicated and cumbersome to accommodate even minimal arithmetic. The widespread adoption of Arabic numerals changed that, providing merchants with a standard, easy-to-read language for expressing levels of raw materials, inventory, sales, and cash reserves.

In his masterful history of accounting, Jacob Soll suggests that reporting within a firm took off when its leaders realized that more information flowing to them could be translated into better control over the activities within their enterprise, as evidenced by the ascent of a number of Italian merchant families—the Medici, the Bardi, the Peruzzi, and others—to the heights of commercial success, establishing them as the preeminent bankers of fifteenth-century Europe.

First among them was Cosimo de' Medici the Elder, who took over the family firm after his father's death, in 1429. When Cosimo inherited the business, double-entry bookkeeping, the idea that every business transaction should be represented twice in an accounting system, as an influx and outflow of value, was no longer a novel idea; in fact, all merchants in Florence were required to maintain such records for the calculation of their tax bills. But bookkeeping was seen as a duty owed to the state, not as a mechanism for internal control. Cosimo turned that around and made bookkeeping into a powerful tool of informational oversight that became an integral part of his firm's daily practice. He insisted on receiving regular, stringently reported information from every branch, which was condensed into a simplified set of books that made it easy for him to detect errors and inconsistencies. The men in charge of each office also had to agree to an annual audit, conducted by Cosimo himself with help from his immediate staff. Cosimo was a shrewd businessman, but his secret lay in understanding that from the comforts of his Florentine mansion, he could use the flow of information generated by accounting to control his financial empire.

Over the next several decades, he expanded the business, setting up dozens of branches and agents across Europe, including offices in faraway Bruges and London. His success did not result, as has long been assumed, from the practice of putting his relatives in charge of these far-flung offices, thereby ensuring the loyalty and honesty of their services through close, and self-serving, familial ties. Rather, the Medici prospered under Cosimo because he retained personal control over the firm's accounts, and thus over key information flows, updating them daily and balancing them frequently.

Such accounting offered several advantages. It required that the "books" be kept balanced at all times; that way, tabulation errors could be caught more easily. Merchants could also keep tabs on their agents—by looking for discrepancies in the data—and discover whether an agent was trying to embezzle monies or hide unpleasant news about the performance of his office. With accounting, the head of a firm could determine whether his agent's actions were successes or failures and act accordingly. There was a material reckoning. But good accounting also laid the groundwork for steady information flows that allowed firms to extend the scale and scope of their operations.

Double-entry bookkeeping did not become the norm in most firms for the next several centuries. The generation of Medici after Cosimo, as well as other wealthy Italians, discarded the practice of accounting in favor of more "intellectual" pursuits, such as politics and the arts. Accounting was deemed to be below their elite status. The modern conception of a company was also still in its infancy, and many executives—including at the storied East India joint-stock companies set up under the auspices of Europe's crowns—were more concerned with share prices and market speculation than with internal control, efficiency, and the steady generation of profits. Most firms might keep a set of basic books and ledgers, balancing them when required by law, but the accounts were often inaccurate; and in some cases, the books were "cooked," obfuscating a company's financial problems in order to keep investors' money rolling in. Unlike Cosimo, many leaders of firms had yet to realize that accounting provided them with a steady stream of important information about the inner workings of their ventures. In part, this was also because the early pioneers in bookkeeping were almost exclusively focused on cash flow: the Medici and other Italian merchants of the day were primarily bankers. They traded merchandise, but mostly they traded money. It wasn't until the eighteenth century, when a young English merchant named Josiah Wedgwood started looking at the costs of production, that the full potential of accounting was realized.

Wedgwood had established a name for himself in pottery. The wares turned out by his factory were coveted by the English aristocracy, including Queen Charlotte, who allowed him to sell patterns branded as Queen's Ware to the general public. Despite this grand coup, Wedgwood wasn't getting rich. He kept good books, so he knew that his expenses for materials and labor were nearly even with his sales revenue. He was barely making a profit, but his ledgers did not tell him why that was the case. So he started dissecting every step of his manufacturing and distribution operations and recorded every cost associated with each step, inventing comprehensive cost accounting in the process. With this information available, he could identify where resources were being wasted, which patterns had too many people working on them, and which products yielded the highest profit margin, so he could continually decrease costs and increase revenues through analysis and allocation.

In short, Wedgwood transformed the information flows from accounting for the past into a strategic tool for business planning, for preparing for the future. Comprehensive cost accounting enhanced the performance of organizations built around a central decision-making authority. In the nineteenth century, when reporting and accounting turned into valuable tools to formalize and standardize information flows within firms, it greatly facilitated the rise of the firm as a highly efficient way for humans to coordinate. The share-issuing corporation was an important innovation to limit risk and accumulate capital, and population growth as well as globalization of trade gave companies a chance to scale; but through comprehensive reporting, a firm's leaders would gain sufficient informational control to navigate a firm on a path to sustainable profit.

Such control, of course, requires more than the right techniques and the relevant ledgers. At least as important is the formation and training of a cadre of individuals within the firm, selected for their honesty and diligence, who are assigned the role of keeping accurate records and reporting comparative results. The accountants are the unsung heroes of the rise of the firm: their tasks are often thankless, at times repetitive, offering limited opportunities for creativity. Indeed, when bookkeepers get creative, attempting to emulate innovation processes elsewhere in the firm, it's rarely to the firm's long-term benefit. The efficiency gains of the firm depend on honest and accurate accounting.

All the reporting in the world has no value, though, if it isn't taken seriously by a firm's top decision makers. For the key concept of the firm—centralizing information flows and decision-making as a tool of comprehensive control—to attain its full potential, the concept needs to be deeply embedded in a firm's inner workings. This kind of extensive reporting commenced in earnest around the 1890s, when American engineer Frederick Winslow Taylor championed a new school of thought. Today, Taylor is mainly remembered for advocating the collection of minute details about every task performed in a factory. Although this was sometimes effective—as at Bethlehem Steel—"Taylorism" was often resented by employees, who felt that quantifying every aspect of human labor turned workers into mere cogs in the industrialists' machines. But Taylor was concerned with much more than speeding up the movements of workers on an assembly line. He argued that comprehensive information flows and processing were essential to a new brand of organizational leadership he called "scientific management," based on reporting, accounting, and, most of all, calculating comparisons. This regimented system, Taylor insisted, could and should be taught to every budding manager. In the throes of mass production, he found an audience eager to codify paths of control leading to efficiency. His ideas were used as the basis of the curriculum for the first master's degree in business administration offered by the Harvard Business School.

New technologies eased the collection and communication of the vast amount of data required by Taylor's scientific management. A generation of executives eagerly enlisted the help of the punch-card tabulator invented by Herman Hollerith, whose company later became IBM. When Hollerith first hawked the invention to companies, they demanded more than a simple counting machine: they needed information that aligned with their corporate goals and improved the bottom line. For one of his earliest clients, the New York Central Railroad, Hollerith created the capacity to add together the numbers recorded in a single field on multiple cards. This allowed the railroad to efficiently audit millions of its waybills—which provided instructions on how much of a shipment had been carried, and how far, on its trains—and made sure that it was charging its customers correctly. But the data could also be used to identify matching shipping customers.

Scientific management has led to a huge increase in information collection and processing activities among firms. Corporate leaders actively aim for better information flows; it's what drives Amazon's Jeff Bezos to require his executives to collect, share, and be held accountable for large amounts of data. But it also requires a firm's managers to process and comprehend these data streams, thereby threatening to overload decision makers. This is the flip side of the huge improvements in the flow of information within a firm.

AS MORE AND MORE DETAILED INFORMATION FLOODS TO the center, the firm must ensure that it is translated into good decisions. This has always been a deeply human task and technical tools have been of limited use. Hence, enhanced decision-making in firms has relied on organizational innovations, such as finding ways to spread the task across more individuals. By devising guidelines for standard decisions, coherence of decision-making can be maintained even if an increasing number of people are involved in it. Choosing and training a firm's leaders to become superior decision makers, too, helps to cope with the need to translate growing streams of information into an increasing number of decisions. These and similar organizational strategies improve decision processes in firms. They are powerful, but they also entail weaknesses: excessive delegation, for instance, reduces coherence. To protect against these weaknesses, further measures need to be taken, resulting in more complex organizational setups that need to be customized to fit a particular firm and its context. In contrast to innovations in reporting, there are no easy, simple wins when it comes to improving the process of decision-making within firms.

Delegation of some decision power down a firm's hierarchy, of course, is the most obvious strategy, not just to spread decision-making to a larger group of people but also to triage decisions, have local decisions resolved locally, and bring only the most important and most general decisions to the top. But delegation of decision-making is a delicate balancing act. If too many decisions are delegated, the decision load at the center is reduced, but so is a firm's effectiveness. On the other hand, delegate too little, and executives are overburdened. Done right, however, delegation paired with the appropriate information flows may work well.

Consider the evolution of General Motors, which started out building carriages for horse-drawn vehicles. In the early 1900s, it was the biggest manufacturer of such vehicles in the United States. The firm's founder, William Durant, saw an opportunity in the struggling Buick Motor Company, which had built one of the earliest internal combustion engines, and decided to buy it. Then he bought another fledgling carmaker, then another. By 1920, GM was a conglomerate to which company after company had been added with little consideration for how they might all fit together or how information might flow to key decision makers other than through Billy Durant himself. Yet despite the company's size and scope, GM's car lines were getting beaten by Ford's ubiquitous Model T. Durant was ignominiously forced out by his investors after they hired an outside consultant to "evaluate the efficiency of General Motors' management" and found that the buck stopped with Durant—all information, decisions, and financial resources flowed through him, leaving the company paralyzed during the severe economic downturn of 1920.

Durant's successor, engineer Alfred P. Sloan, was a born organizer and immediately began improving the efficiency of the company's operations. But instead of tinkering with the assembly line, Sloan focused his efforts on optimizing the flow of information. He consolidated several central functions, such as accounting, to improve data gathering. Because dealer sales reports and projections took too long to make their way up the corporate hierarchy to Sloan's office, he hired an outside firm to collect the data.

Sloan organized the unwieldy corporation into various divisions, each focused on a unique segment of the market, and he delegated decision-making authority to the executives in charge of each division. This vastly streamlined GM's information processing, because managers in each division only had to digest data about their particular segment of the market.

Soon GM had overtaken Ford as America's biggest carmaker. Sloan's *My Years with General Motors* became a management bible. The firm's financial controls "supplied clear, standardized, and frequent performance information," according to a study in the *Harvard Business Review.* "Armed with these tools, executives could make decisions with up-to-date information, reducing the influence of personal loyalties or entrenched perspectives." Improving the flow of information while reducing decision-making bottlenecks through delegation was the combination that led to GM's success.

After World War II, Ford set out to emulate this strategy, hiring a series of data "whiz kids." Robert McNamara—once called "an IBM machine with legs"—who had saved the US military billions of dollars in procurement costs during the war, was one, and he would eventually be named Ford's president. McNamara brought modern information management to Ford. Fully aware of our human cognitive limitations in comparison to Herman Hollerith's punch cards, McNamara throughout his tenure at Ford, and later as secretary of defense and president of the World Bank, combated the inability of the human brain to process large amounts of information. He ruthlessly reduced complexity, simplifying data to make it more digestible to the human mind, much as market participants focus on price.

McNamara's efforts sometimes were hindered by the lack of real-time, comprehensive data. Employee reports and punch cards had to be read and analyzed. Simplified metrics did not capture the entire picture. Thanks to digital computing, by the 1970s it became increasingly possible to consolidate the many different data sources in firms that cover everything from financial accounting and reporting to human resources, manufacturing, inventory management, and sales into a single database. Called enterprise resource planning (ERP), and pioneered by German start-up SAP, these systems gave managers tools to shape the flow of information much more comprehensively than ever before.

The big challenge when delegating decision-making is to ensure that the advantages of centralized information flows and decision-making, such as reduced information needs across the organization as well as coherence of decisions, aren't lost. Newly empowered decision makers, often located away from a firm's center, must be trained to make choices that align with central leadership. If they don't, the "noise" they create in the organization causes inefficiencies.

There are several time-honored management strategies to avoid such "noise," including the establishment of standard operating procedure (SOP) and other rule books that guide decision-making in firms. To some extent, these rules serve to communicate from the top down by giving employees clear, comprehensive instructions on how to make decisions in a set of identified scenarios. This ensures that the leaders' decisions are applied consistently throughout the organization. But equally, such standardization aims to lower the volume of decisions that must be presented to the firm's leadership. By laying down such general rules, delegation is facilitated. A popular information-processing tool, the checklist—which is essentially a simplified version of the

SOP—provides similar decision-making benefits. Checklists in aircraft have been shown to significantly reduce pilot error, particularly in high-stress situations. And Harvard Medical School professor Atul Gawande has demonstrated that using a checklist in a hospital intensive care unit drastically lowers infection rates among patients. Checking all the boxes may be tedious, but it makes for better decisions. Checklists and SOPs work well in routine situations—those that have largely been anticipated in advance—and when the outcomes are the same, or nearly so, each time.

Some firms, especially those in which employee turnover is relatively high, take the strategy of SOP a step further, embedding some of the rules and procedures into the firm's tools and technology. For example, if a protective shield must be closed manually before a welding machine can be operated safely, it might be better to link a sensor to the machine and prevent it from welding unless the sensor confirms that the shield is closed. As machines grow more sophisticated, the rules that can be encoded increase in complexity, too. But even more so than with SOPs, the ability to delegate decision-making is limited by predictability. Machine-based rules only work for very routine tasks that have been studied in detail.

There are countless variations on these information-processing strategies, including far less rigid ones. In every case, the firm's leaders adopt them in order to shift some decision-making power down the hierarchy. They hope to strike the right balance between reducing the information burden on themselves while maintaining control and consistency across the organization. This is what Alfred P. Sloan mastered at GM: balancing centralized and delegated decision. Sloan called it "decentralization with coordinated control."

By the end of the twentieth century, decision makers in firms had more information at their fingertips than ever. It was an astonishing leap forward that began with the practice of keeping good books more than half a millennium earlier. The need to create and maintain comprehensive information flows had become a basic tenet of good management. And as the case of Amazon underscores, even digital darlings are sticking to tried and trusted concepts for managing information flows.

Of course, incorrect and incomplete information still exists in firms. Sometimes even some of the most important information fails to travel to decision makers when it should, because of fear, groupthink, or in some cases, perhaps because nobody cared. But by and large, the firm has become a highly effective way for humans to coordinate efficiently. It has gained significant ground relative to the market as the coordination mechanism of choice. But much of the efficiency gains have now been absorbed, and to reap further benefits firms will have to focus on improving information processing among a firm's top decision makers. There, however, firms face a far more difficult challenge: constraints in human cognition.

As psychologists Daniel Kahneman and Amos Tversky pointed out in their groundbreaking studies (which fueled the creation of a new academic field, behavioral economics), humans are plagued by a range of fundamental cognitive limitations that impair our general ability to decide well. We have seen this in the context of price, but the constraints are far more universal. For example, we naturally evaluate new information by comparing it to information that we can easily bring to mind. This results in our overestimating the probability of dramatic events, such as plane crashes. We are also inclined to discount the benefits of making a change as compared to sticking with the status quo, preferring the known to the unknown. Further challenging the firm's ability to improve decision-making at the top, humans aren't afflicted with these biases exactly the same way. Some of us are more prone to some biases than others. In markets, the variety and diversity of decision makers tend to mitigate these cognitive limitations (but by no means eliminate them). By contrast, in firms, the centralization of decision-making amplifies these cognitive limitations.

To improve decision-making in the firm, one would have to improve the cognitive capabilities of its decision makers. Is this possible? Perhaps, some may suggest, companies could select people for leadership positions who are less likely to fall prey to cognitive biases or similar errors in decision-making. There is indeed evidence that some of us may be better at some aspects of assessing information than others. Studies have shown that men are more likely than women to exhibit *confirmation bias*—seeking out or putting more weight on information that confirms a preexisting belief. People from Western cultures are more prone than people from Asian cultures to the *fundamental attribution error*—believing that others' performance and behavior derive from their personalities and temperaments rather than from the larger culture or environment. But these relative disadvantages only seem to assert themselves with respect to a single bias. There is also no direct relationship between intelligence and cognitive biases. At least in this context, being smarter doesn't necessarily lead to being better at making decisions. Thus, there is no foolproof method for choosing a leader who will magically overcome cognitive shortcomings. Through selection, we cannot escape what the political scientist Herbert

Simon called our bounded rationality—the limits of our ability to make optimal decisions.

Another strategy would be to hire a group of leaders whose cognitive strengths complement each other so well that the biases of the team as a whole are reduced. But even if this were possible, it's not clear whether such cognitive strengths wouldn't lead to deficits in other areas of responsibility. After all, a firm's leaders are also charged with inspiring and motivating employees, developing products and services, and communicating with customers and shareholders. Charisma, expertise, and energy are just as important for leading a firm as decision-making aptitude. A selection process focused largely on a person's skill in information analysis would likely affect his or her performance in these other roles.

Some firms have looked for ways to improve the decision-making aptitude of their executives up and down the ranks, but with limited success. They hire and promote graduates of business schools and management training programs that promise to teach students how to translate information into decisions and be less prone to cognitive biases. There are studies that show that unambiguous feedback about a decision—something as unmistakable as a pilot learning that if the wing flaps are not extended, a jetliner will not get enough lift to take off or a doctor seeing that administering a particular drug causes a patient's blood pressure to fall—can decrease cognitive bias. But few management decisions result in this kind of unequivocal feedback.

An alternative view of how to improve human decision-making has gained traction in the past decade. It asserts that humans aren't such bad information processors, as long as they follow their intuition and adhere to a few fairly simple heuristics, or shortcuts developed through trial and error. Proponents of heuristics, including Gerd Gigerenzer, director of the Max Planck Institute for Human Development, do not dispute the extensive evidence for the existence of decision biases. They argue, instead, that the carefully constructed experiments of behavioral economists are removed from reality.

That's a reasonable criticism. Almost all scientific experiments aim to reduce the number of variables involved in order to tease out cause and effect, and many psychological experiments present subjects with fairly binary choices. But life is rarely so starkly black-and-white. According to Gigerenzer, humans have already developed information filters to "protect us from some of the dangers of possessing too much information." The answer, he says, is not to find ways to make up for our cognitive limitations but to rely on the "gut feelings" that we have successfully utilized for millennia.

If you're an executive who has to make many important but quite different decisions every single day, this may sound like quite an enticing option. Forget about analyzing information to get a deeper, richer, more detailed understanding of what's going on. Disregard the complexities of the world and celebrate simplicity—informational ignorance works just as well, and relying on your instincts is cheaper and faster than dealing with information overload. It should come as no surprise, then, that Gigerenzer's advice has been embraced by quite a few corporate leaders. Around three-fourths of business executives, according to a 2014 survey, trust their intuition about decisions. Even two-thirds of those who describe themselves as "data-driven" professionals—they "collect and analyze as much [information] as possible before making a decision"—said that they value their gut feelings. Who wouldn't prefer the cheap and simple option to hours spent analyzing information and translating it into decisions? Who wouldn't want to succeed without having to try hard?

The fundamental problem with heuristics is the same as the problem with medieval bloodletting. It may be okay as long as one has no better alternative—if one can't gather information comprehensively, process it, and translate it into appropriate decisions. But just as babies give up crawling when they learn to walk, eventually every second-best solution to a problem is cast aside for the real thing. If we've learned anything from the Enlightenment, it's that ignorance never is bliss, only a temporary fix and a reminder to try harder. So heuristics don't offer us a simple solution to the cognitive decision-making challenge in firms.

OVER THE PAST CENTURIES, FIRMS HAVE IMPROVED THEIR internal information flows, turning them into a powerful means of organizational innovation. This has conveyed a competitive edge for firms over money-based markets and their information reductionism. Translating information into decisions hasn't progressed as much, not because firms haven't tried, but because of constraints in human cognition. There's only so much that delegation, standardization (through SOPs and so on), and selection (of the right employees), and even heuristics can yield. As the low-hanging fruits of information flows have been harvested, further progress needs to come from improved decision-making, making it much more challenging. The real reason that Jeff Bezos chose to mold Amazon as a fairly conventional firm is an indicator of how much of a challenge that is. Even though we are now well into the digital age, for Amazon a traditional hierarchical setup has turned out to be the most efficient choice.

Data-richness fundamentally disrupts this situation. Digital tools that may have been used to speed up what we've been doing may also enable us to reorganize the social mechanism of coordination. Using cutting-edge data-focused tools, market participants can greatly enhance their ability to find optimal matches. This tips the competitive balance between markets and firms in markets' favor. As markets evolve, are traditional firms the big losers in the shift from money-based to data-rich markets? Couldn't firms utilize the same data-driven tools to handle their information overload, ease the crushing quantity of decisions, upgrade executive decision-making, and benefit from the data age just as the market does?

Not to the same extent as markets, as we discuss in the following chapter. That doesn't mean that firms have no future, but it may prompt many of them to reevaluate how they are organized. Traditional firms that are feeling mounting pressure from digital disrupters may be the first to change. Over time, though, even Jeff Bezos may have to rethink how to best run Amazon.

– 6 –

FIRM FUTURES

DECEMBER 26, 2016, BORE BAD NEWS FOR INSURANCE claims assessors, even though few of them noticed. The worrisome information was buried deep in a press release issued by Fukoku Mutual Life Insurance, a Japanese giant headquartered in downtown Tokyo. The company announced it would use Watson, a machine learning system developed by IBM to assess insurance claims. It optimistically asserted that with this measure, "the burden of work processing can be reduced by about 30 percent." It was left to one of Japan's most prominent newspapers, the *Mainichi Shimbun*, to report a few days later that hiring Watson would make around one-third of the staff in Fukoku's claims-assessment department redundant.

Some months earlier, halfway around the world, in Stuttgart, Germany, employees of the automobile manufacturer Daimler were asked to give up their customary offices—but for a very different reason. Daimler CEO Dieter Zetsche announced a radical restructuring of his company, a symbol of German corporate conservatism, and its traditional top-down management culture. His goal was to have 60,000 employees—around 20 percent of Daimler's global workforce—operate outside their former reporting lines and corporate silos within one year. Knowledge workers at Mercedes-Benz's parent company weren't laid off; they were asked to become part of a reshaped organization with flexible teams and less hierarchy.

Zetsche is a tall, slim man with a background in electrical engineering. He is easy to recognize, thanks to his bushy mustache and his sense of humor. Confronted with potential disruption by new competitors such as Tesla and its Chinese counterparts, facing game-changing advances in technology, such as self-driving cars, and finding himself up against novel business models, such as ride-hailing services, the Daimler head ordered his organization to shift into start-up mode. Copying the cellular organizational structure of successful Internet titans such as Google, Zetsche explained, would mean that Daimler will "supplement the hierarchical-management pyramid with cross-functional and interdisciplinary groups and eventually replace them." The goal is to encourage faster innovation, and the key to achieving this goal is drastically improved decision-making: "At the moment, we have up to six levels of decision-making. We want to have only two decision-making levels for every issue by 2020."

Fukoku and Daimler represent two very different responses to the data revolution that is empowering the market. But their strategies share the same goal: to save the firm. Neither company is acting in panic. Their mainline businesses are in good shape. Both old-economy incumbents realize, however, perhaps earlier than some of their competitors, that when the market changes fundamentally, so must the firm.

The Japanese insurer's strategy centers on automation, using data-driven machine learning systems to make decisions that formerly were made by white-collar workers. The German carmaker takes another tack: streamlining managerial decision-making processes (while continuing the rapid automation of car production). Both strategies aim not only to hold old and new competitors in check but also to defend the firm against an improved, emboldened market. Intriguingly, Daimler's weapon of choice is taken from the market's own arsenal. But will it work?

FIRMS EXIST WHEREVER THEY CAN ORGANIZE HUMAN activity more efficiently than the market can. Firms persist when they operate more efficiently than the firms they compete against. The most obvious strategy for firms in response to the renaissance of the market, therefore, is to improve their own efficiency. This option lies at the heart of Fukoku's decision to employ Watson.

It's the same rationale that, from time immemorial, has led firms to supplant humans with machines, from the steam engine to the factory robot. The strategy predates the Industrial Revolution. And it was well before the windmill replaced humans grinding grains and the printing press supplanted scribes. In fact, the ancient invention (or perhaps discovery) of the wheel itself enabled cargoes to be transported much more efficiently than through human carriage. The strategy to favor machines over humans gathered speed with the introduction of John Kay's flying shuttle and James Watt's steam engine in the eighteenth century.

Initially, machines often were only marginally more efficient. They required considerable institutional support (such as appropriate laws, financial instruments, and the like), but improvements over time (including organizational changes) increased their efficiency by leaps and bounds, and they ultimately became indispensable in virtually all sectors. Similar efficiency gains were realized through the optimization of every process within a firm, whether it was called Taylorism, Six Sigma, or lean management.

Whether a focus on efficiency is a suitable strategy largely depends on two factors. First, are there existing inefficiencies within an industry that can be eliminated? The traditional network airline business model, for example, was wasteful enough that more efficient, low-cost airlines could take over a substantial share of the air-travel market. By contrast, modern large supermarket chains are so comparatively efficient in their operations that new entrants (including digital start-ups— remember Webvan?) have had much less opportunity to disrupt them. Of course, it's important to understand not only whether a given industry sector has efficiency opportunities but also whether a particular firm, by focusing on efficiency, can improve its relative position. Most traditional airlines may recognize their inefficiencies and see the advantages of the low-cost carrier model exemplified by JetBlue and Ryanair, but they lack the ability to reform themselves due to a variety of factors, from existing obligations to ossified organizational structures.

The second important factor is time. Relative efficiency improvements almost always result in an advantage that's only temporary as other firms catch up. That's why firms competing on commodity products often start a new round of economizing the moment they conclude the previous one. Over time, as inefficiencies are eliminated, there's a diminishing marginal utility to pushing for automation, process efficiencies, and other cost-cutting measures. They are only infrequently successful in the long term—but then again, most business strategies in general are rarely successful over the long term. Often the secret to success is continuous adaptation to the environment and the ability to choose the right strategy in any given context.

Keeping these constraints in mind, Fukoku's shift to Watson is a quite conventional choice. But that's okay, because the insurance sector is relatively inefficient: a lot of paper gets processed and pushed around; it's labor intensive and has seen limited automation. Profitability depends on setting premiums at the right level and assessing claims efficiently, but these decisions, although fairly straightforward, weren't standardized enough to have been easily automated until sophisticated data-driven machines like Watson arrived. In that sense, insurance today is like steel production before the invention of Watt's steam engine.

Fukoku's strategy to start automating these clerical decision processes is likely to pay off nicely. Installing Watson cost the company $1.7 million in up-front investment. Yearly maintenance is estimated to be around $130,000, and the company aims to save roughly $1.1 million per year on salaries. Their return on investment within two years is much faster than for heavy machine-based automation in manufacturing. Fukoku concedes that experienced employees will still have to check and approve the decisions the new system will suggest, but Fukoku is already looking into installing another machine learning system to do this double-checking.

For decades, tech pundits have claimed that artificial intelligence will replace human knowledge workers. It always seemed possible, but it has rarely happened. Even routine clerical decisions at insurance companies weren't standardized and simple enough to automate with artificial intelligence systems based on fixed rules. But artificial intelligence has evolved—from systems based on general rules to learning ones—trained on massive amounts of data. With Fukoku's switch to Watson, automated decision-making for routine claims has reached its Kitty Hawk moment. Just as it did in the realm of powered flight, it has taken ages to reach the moment of takeoff. Having mastered the technical underpinnings, it has reached its tipping point and the sky is no longer the limit. From here on, automation of routine decisions, not just in insurance but in every sector, will be in full swing. Much as Henry Ford cut costs by automating the car-manufacturing process, Fukoku cut costs by automating the insurance-claims process: different sectors, same idea. And, like Ford, Fukoku will reap an efficiency dividend from automation, thereby improving its competitive position.

Some may ask, why stop there? Why does automation have to end with office clerks? Why can't it extend to middle managers and even the C suite? It's an intriguing idea: use machine systems to assist managerial decisions and bump artificial intelligence into the executive ranks.

Bridgewater Associates, the world's largest hedge fund, is about to do exactly that. It plans to build learning systems that will not only choose investment opportunities for its $160 billion in assets, as it has been doing for years, but will also make general management decisions, such as whom to hire, whom to promote, and whom to fire. Ray Dalio, Bridgewater's legendary founder and CEO, is known for his passion for data, and he aims to automate three-fourths of all management decisions by 2022. Bridgewater is better positioned than most companies to build such systems because of how it handles data. Although firms do not routinely capture their decision-making processes as data (at best they record the outcome of these processes), Bridgewater is constantly adding to its rich data streams. For example, most meetings and business conversations are already logged for future data mining; and employees are continually asked to rate their colleagues' performance.

To make sense of all this data, Dalio hired David Ferrucci, the developer who led Watson to its *Jeopardy!* victory in 2011. Ferrucci heads a team at Bridgewater called the "Systemized Intelligence Lab." Its main task is to develop a comprehensive management-software system called PriOS that will drive decision-making at Bridgewater. In some contexts, PriOS will augment rather than replace human decision-making, but the ultimate goal is for humans to be relegated to just setting the criteria the system uses in its decision-making.

It's hardly surprising that artificial intelligence has been readily accepted in the data-savvy and algorithm-friendly domain of investment banking. But the trend is spreading to other, less obvious sectors, including traditional human services. Imagine, for example, a machine learning system making strategic decisions in a company focused on caregiving for the elderly. A robo-boss with enough data at its fingertips could run a large team of caregivers predictably and efficiently: it could apply a holistic approach to planning weekend shifts, never forgetting to budget extra time for some patients, and could swiftly locate and notify a nearby caregiver in case of an emergency. A robo-boss would never favor one employee or one client over another out of personal sympathy, making decisions less biased.

Or think about such a learning decision-assistance system at a small construction company that is considering whether to buy, lease, or just rent an additional backhoe. Its founder, Bob, would typically make the decision based on his limited supply of information and on his gut feelings. He might be biased because he loves the look of the shiny new backhoe, or he might fail to see its value beyond daily rental fees. In contrast, a machine learning system would aggregate data from the company's order book and accounting system, add information about the backhoe's predicted maintenance cost and local economic trends, and conclude: "I'm sorry, Bob. I'm afraid I can't support your buying this fantastic Caterpillar machine, although I know it has excellent performance ratings and great reviews in online forums for equipment nerds."

With data-driven machine learning systems up and running, we could end up with thriving firms staffed by fewer managers. Automation would finally have reached every corner of a firm, including its decision-making leadership. But a fundamental difficulty in pushing decision automation far into management territory remains.

Straight transaction decisions on markets are all fairly similar: we compare our preferences with those of others and then decide whether and with whom to transact. The specifics may vary, but market participants encounter decisions of this kind every time they engage in the market. Executives in firms, by contrast, face a wide variety of different *types* of decisions. Thus, on data-rich markets, where comprehensive preference data is becoming available, we can use the data to train decision-assistance systems. We don't have the same luxury in firms, where decisions of a similar type are far fewer, so the data available for a particular type of decision is limited.

As executives face a more varied set of decision options than market participants (the latter "only" have to decide whether to transact), more data is needed for machine learning systems to work, not less. Worse, even where sufficient information inputs are available, data about the actual process of decision-making is still lacking. Without knowing how various inputs were evaluated, weighed, compared, and then translated into a decision, machine learning is stunted. In short, the firm's comparatively limited trickle of relevant data does not rival the market's data flood. Of course, firms could aim to gather more such data, but given the diversity of decisions to be made, it is hard to imagine machine learning systems taking over general managerial decision-making from humans anytime soon.

Could firms rejig their organizational structure to concentrate similar decisions in one unit, resulting in sufficient data for machine learning systems to work? Fukoku could shift to Watson precisely because the company had one unit that was assessing similar claims. There was enough data available for Fukoku to put Watson in charge. If routine managerial decision-making isn't so different from routine clerical decision-making, could machine learning systems do both as long as there is sufficient data? This isn't so different from the old idea of delegation—by dividing up tasks, specific expertise can accrue over time that improves decision-making efficiency. So in an increasing number of organizational units data-driven managerial decision automation may seem to be the logical next step.

There is certainly some movement in that direction. Providers of data-driven machine learning systems such as Daisy Intelligence are already offering retailers a comprehensive tool kit of decision-assistance systems that help in managing inventory, choosing weekly special offers, placing products on the selling floor, and pricing them optimally. It surely won't work for small mom-and-pop shops or fashion boutiques with a niche clientele, because those types of stores don't generate enough data, but it may be just what the doctor ordered for a midsize retailer selling thousands of items, facing strong competition, and having to make fairly similar decisions about inventory and price. The promise associated with such systems is not only that they will improve the quality of routine managerial decisions but also that they will do so at a lower cost. Managerial decisions that are low-level, routine, and expertise-driven, with few interdependencies to other parts of a firm, and supported by sufficient data will be automated; and managers will be replaced by data-driven machine learning systems.

By the same token, managerial decisions that aren't supported by enough data, aren't routine, or require a lot of coordination with other parts of the company will remain largely immune from automation. It is still easier for individuals to share information and thus coordinate with others informally—for example, over drinks after work—than it is for a bot to do so. A machine learning system that uses only data from a single unit may yield a fine decision based on that data, but it might be a bad decision for the firm as a whole.

Human managers have fewer difficulties navigating complex decision-situations involving multiple organizational units. They often combine proficiency in their own units with an ability to communicate and work with colleagues in other units. These managers, who have what's called a T-shaped skill set, blend in-depth expertise in a specific field (represented by the vertical stem in a T) with an ability to collaborate with managers in other areas (the horizontal bar on the T). Machine learning systems may acquire vertical expertise, but their very design prevents them from becoming horizontal collaborators. And without decision makers who are able to see beyond their own units, firms will break apart and fail. That is reassuring news for humans: perhaps we may still be irreplaceable as top-level decision makers in firms, at least until machine learning systems being trained with lots of data from lots of different decisions attain general-purpose decision-making skills that so far are our exclusive domain.

It offers more than reassurance, though. Understanding the limitations of machine learning systems also clarifies the skill set of a successful manager in the medium term: rather than banking on deep expertise in just a single area, the effective corporate leader of tomorrow will be much closer to a classic renaissance person. Such a manager will have a fair amount of knowledge in many areas, thus becoming empowered to contextualize information and see the big picture, the forest rather than just the trees. Moreover, future leaders will need skills beyond those that are directly linked to decision-making. This includes collaboration—the ability to foster interdepartmental and interdisciplinary action.

Finally, and perhaps most important, human managers are needed to facilitate radical innovation—not the kind of continuous improvement that data-driven machine learning offers but the fresh new ideas that aren't contained in data. Lacking human imagination, current artificial intelligence systems have no point of reference against which to navigate the utterly unknown. Of course, this too may change eventually. Perhaps we'll realize that radical innovation isn't so radical after all and is more an extension of existing ideas, or that radical innovation can be emulated through the introduction of randomness. Machine learning system experts around the world are actively pursuing strategies to teach their systems to be creative, albeit currently with only very limited success. It seems that so far, creativity is a difficult nut to crack for machines. Until this changes, managers will continue to be needed to steer the complex process of creative destruction, which famed Austrian

American economist Joseph Schumpeter saw as the source of sweeping innovation.

FUKOKU'S CHOICE—WHICH WE CALL OPTION ONE—focuses on cost. This solution works only if firms successfully automate and optimize decision-making, cutting down on overhead and permanently reducing the size of the workforce. But at its heart, option one is a bet on the past, a bet on a strategy that will disrupt its workers rather than its structure. Fukoku may look like a highly digital firm from the outside because it relies on data and the latest digital tools. But although these tools replace human labor with machines, the internal structure of the firm—its hierarchies and departmental silos—remains unaffected. For management decisions, data will continue to flow in centralized fashion to the firm's leadership. Renaissance managers may thrive there, but, overall, highly traditional command-and-control environments will endure. The most striking difference may be that machines are better than humans at obeying executive orders.

Daimler's strategy—which we call option two—is different. Rather than focusing on machines, the upmarket car manufacturer is rewiring managerial decision-making processes to gain an advantage not just relative to competitors such as BMW and Tesla but relative to the market. Organizationally, it seems like a bet on the future: Daimler is willing to sacrifice historic hierarchies to keep pace with a reinvigorated market and digital challengers.

Daimler's challenge isn't unique; in fact, it is typical for a large, successful corporation. To ease the burden of decision-making, firms have aimed to find the optimal balance between the advantages of decision-making at the top and the need for some decentralization to prevent decision overload. It's a delicate undertaking—and firms that believe they have found their decision-making sweet spot often try to hold on to it at all costs. Decision-making processes that seem to be in balance because they once worked, may in fact have steadily moved away from being optimal. Holding on to old solutions then will lead to an organizational inability to decide, to adapt, and to evolve in time.

Dieter Zetsche's call for a fundamental change is aimed at Daimler's own ossified management structures and fossilized decision-making processes. He complains that decisions take too long, leading to inertia and a failure to adjust swiftly enough to changes in the competitive environment. Zetsche also wants to see more innovation—more ideas that break traditional molds—across Daimler's huge global footprint. His stated goal is to increase responsibility and ownership among decision makers. Zetsche isn't suggesting a new wave of automation; rather, he's tinkering with the organizational setup of his firm. He wants to shift it toward a simpler, faster, flatter model in which decision-making is decentralized.

This is also the mantra of Spotify, a recent digital darling. The Swedish start-up operates the biggest music-streaming service in the world—more than 100 million music lovers use it to listen to their favorite songs and albums. Spotify's business model disrupted Apple's digital music shop, iTunes, which itself was the disrupter of the music industry's longtime business model of selling physical records and CDs in brick-and-mortar shops. The music industry both loathed and loved iTunes. On the one hand, Apple's digital content store broke off a huge chunk of its profits. On the other hand, iTunes shared revenues with content producers, unlike its predecessors that had indulged in illegal peer-to-peer file sharing.

Music streaming was a more fundamental disruption than online music stores because it created a platform that dissociated price from product, cost from consumption. Most users enjoy the service for free: rather than paying to listen they supply personal data (i.e., they accept advertising targeted specifically to them). As of 2017, the 50 million "premium" users dole out a modest all-inclusive monthly fee of $10 for unlimited access to more than 30 million songs, and no one is penalized for listening to music constantly or for skipping tracks a lot.

Nor is there a benefit to frugality. Unlike conventional marketplaces, price has almost completely lost its traditional informational function on Spotify's music market. It is being replaced by a variety of different signals such as information about what music is searched for, what track is skipped and by whom, and what is being shared with friends. Spotify has turned a half-disrupted pipeline business into a matchmaking platform (matching listeners and music, as well as between advertisers and audience) on which direct payments play only a minor informational role.

But what makes Spotify so intriguing for us is its organizational structure. The company was founded in 2006 and launched its portal two years later. Its founders, Swedish entrepreneurs Daniel Ek and Martin Lorentzon, weren't the only ones wanting to see music streaming become a reality. A fierce competition between several streaming start-ups commenced. But unlike some of his competitors, Ek not only happened to be the former CEO of μTorrent, a company known for making file-sharing convenient, he also had a clear vision of how to manage a fast-growing company full of confident coders in the context of an egalitarian Swedish business culture.

Ek has been described as the Tom Sawyer of the lively Stockholm start-up scene: he finds fences and lets others paint them. But the real trick is that he lets others paint them the way they see fit. Spotify's management culture is pretty much the exact opposite of a command-and-control environment. Ek borrowed the principles of agile software development and integrated them in a management system he calls squadification.

"Squads" are small teams responsible entirely for certain aspects of a product (such as the search function or the user interface) or certain business activities (such as sales in a given market). A squad doesn't have a boss, only a so-called product owner, whose task is to ensure that all the members of the team have everything they need to do a good job. The product owner also keeps an eye on the self-defined goals and deadlines of the team, but unlike traditional team leaders, this one has no enforcement power. Neither does the squad's "agile coach," whose task is to foster and facilitate collaboration within the team.

The underlying philosophy is starkly different from that of traditional hierarchical firms: don't ask your manager, because you don't really have one. Experiment with data to create evidence. Then share it with your teammates, the teams working on connected issues, and anyone in the company you consider knowledgeable on a given matter. Get feedback, then make the decision yourself (or as a squad) and implement it as swiftly as possible. If there are problems, it is up to you to fix them.

To avoid silos and foster collaboration across squads, specialists with connected expertise meet in "chapters." Squads working on similar projects form "tribes," which should have no more than 150 members. Above the tribe level, there are "guilds." The main goal of this structure is to facilitate the flow of information and knowledge throughout the company. To ensure coherence of business strategy and decision-making, Spotify's fluid organizational chart denotes the ranks of "system owner" and "chief architect." But in practice even those two top management positions do not have the power to give orders. Much like product owners and agile coaches on the squad level, they operate like moderators. They use the soft power of persuasion and the hard facts of data-driven evidence if they want to see their opinions turned into action.

Spotify prides itself on its culture of feedback and coaching. It's okay to fail as long as people learn from it (in fact some squads even have "fail walls" to highlight the lessons to be learned). Central to this culture is the separation of feedback and learning from formal discussions of salary and performance, which remain linked to a more formal hierarchy. Although peer input is actively sought for learning, an individual's salary and performance matters aren't discussed by the entire squad. Spotify believes this facilitates open and robust feedback, as there is no longer an incentive for people to give each other favorable reviews. The company has even developed an internal tool to elicit frequent feedback from others.

Spotify's squadification model is an outgrowth of the ideals and values that resonate with its founders and its workforce. In part, it works because Spotify isn't operating in a heavily regulated sector such as health care or banking, but in a field where competition is harsh and constant innovation is crucial. And in part, Spotify's organizational setup is well suited for a digital start-up, an environment where data is abundant and can easily be made available.

But there is still another reason that Spotify chose the squadification approach. Once we push aside the communitarian language and the seeming affinity for small, adaptive, loosely organized groups, we see that the company is injecting a piece of market DNA into its structure. After all, decentralized decision-making *is* the hallmark of the market. Introducing such radical decentralization into a firm invites bits of the market inside. Spotify is an intriguing example of a firm intent on avoiding turning into a traditional firm by opting to be an organizational hybrid: part firm, part market.

Introducing the market into the firm isn't a radically new idea. There are many variations, but what unites them all is the delegation of some managerial decision-making to fundamentally decentralized, more market-like mechanisms. John Deere's transition from tractor manufacturer to a world champion of connected farming was possible only as its corporate culture shifted toward faster, more decentralized decision-making in self-organized teams. General Electric and Siemens are decentralizing their supply-chain and production operations into regional units because central decision-making happens too far off, far too often. And media giant Thomson Reuters aims to speed up innovation by launching idea contests that don't just award prizes but also turn the winning ideas into reality by financing them through an internal venture-capital fund.

As the market gains renewed competitive momentum, the firm will find itself in need of catching up. We'll likely see more market DNA, more decentralization, more internal competition in the firms of the future. But as companies aim to embrace some of the market's essential qualities, we suggest firms may want to adjust their market-embracing strategies. As the market itself is changing from one greased by money to one fueled by rich data streams, it may be more advantageous for firms to embrace data-rich markets than to compete with them. The key is to understand that these new markets have multiple components.

Rich information streams need to be augmented by matching algorithms and machine learning systems. When combined in a market context, these can assist a firm's decision makers much like they assist market participants in identifying the optimal match, given multiple preferences. Intriguingly, we are already seeing such strategies being implemented in firms—at least partially—for a very human purpose: internal talent management.

In a 2016 survey conducted by the professional services firm Deloitte, 39 percent of large-company executives said they were either "barely able" or "unable" to recruit the talent their firms required. With such difficulty finding talent outside one's organization, it may be better to look for the right individuals internally. To this end, many large corporations have been improving talent management with data-driven internal marketplaces. It might look like a simple solution, but it addresses a number of the human-resources challenges that firms routinely face.

Firms with a central HR department pool their talent-acquisition and talent-retention expertise into one organizational unit, so that best practices can be shared and, if necessary, adjusted quickly. But like any centralization, this creates an information-processing bottleneck. HR staff may be knowledgeable, but they will have to contend with a lot of information as they optimize talent allocation throughout the organization. Their decisions will require comparing individuals and open positions and matching them across multiple dimensions such as experience level, specific expertise, or salary band. That is challenging for any manager, even if they are competent, hardworking, and well-meaning.

Internal reallocation of talent also faces another difficulty: managers who have a great and promising young assistant will do much to keep that talent in their unit. There is an incentive for managers to underreport the positive qualities of the most competent staff to HR, lest HR be tempted to move one's valued associate into a different position.

To put an end to such inefficiencies, firms such as American Express, AT&T, and IBM have phased in software platforms that go far beyond classified-ad-type announcements of open positions on the company's intranet. They match detailed (albeit standardized) job descriptions with detailed (albeit standardized) talent profiles. Filters make individuals and position pools easy to search, both for employees seeking a new challenge and for managers looking for new talent. And recommendation engines facilitate matchmaking across multiple dimensions.

These internal talent marketplaces offer a number of advantages. First, they decentralize matching, reducing information overload within HR departments. Searching and matching is done outside HR, by managers with positions to fill and employees interested in making a move. Thanks to multidimensional information streams and talent-matching software, the costs of the search are kept comparatively low. And neither side has much of an incentive to underreport, conceal preferences, or exaggerate need. This ensures that information available on the market is fairly accurate and comprehensive.

No longer do managers "own" their talent; no longer can they (ab)use information to restrict the optimal matching of people to positions. As a result, there is fluidity of human labor inside an organization, which increases a firm's ability to adjust to changes swiftly. It also empowers employees to act more like free agents (even if executives will want to retain the power to assign people if necessary). This fits well with the preferences of young people, who are usually eager to stretch their talent muscles themselves, "own" the way they work, shift to new challenges frequently, and avoid getting stuck in one particular career track.

Internal talent marketplaces also have a disincentive to focus on price. In part, this is because firms do not want to institute a mechanism for open internal price competition. In part, though, it is also an attempt to refocus those participating in the market on the diversity of preferences that influence job choices. By using salary prominently in internal talent markets, managers and employees would be tempted to optimize largely on price, thereby running the risk that market participants will overlook other dimensions that are important for job satisfaction. Price can be deemphasized in such internal job-matching platforms by mandating that all parties stick to firmwide salary bands, which are not individually negotiable. That way, some flexibility in agreeing on a salary is retained but the emphasis on it is reduced.

Talent-matching platforms are an intriguing way to inject markets rich with information into the organizational structure of the firm. Phased in about a decade ago, they are now slowly opening the doors to outside talent—in most cases pre-approved freelancers. Consumer goods giant Procter & Gamble has even opened its internal platform, connecting employees involved in innovation activities to individuals outside the organization.

There is room for improvement, applying the concept of data-rich markets that we explained in Chapter 4: information ontologies (hierarchies of keywords) need to be further standardized, and matching algorithms need to become more sophisticated. Most important, these talent markets currently lack machine learning systems, which grasp decision makers' preferences just by observing them. Over the next years, we will see a steady increase in these systems being deployed in companies around the world. It will begin in large corporations with lots of internal matching requests that generate sufficient data—a necessary condition for such systems to work well. But over time, midsize firms, too, will join the fray. While we may initially see a lot of adoption in the HR field, the application of such a multidimensional market approach is by no means limited to matters of staff allocation. We will see it pop up in other areas, such as marketing, procurement, inventory management, finance, even product development.

Once such a market mechanism is in place, its scope may extend outside the firm. Why restrict supply and demand when the mechanism could attract participants from outside the organization as well? Of course there are limits, and some internal markets may never be open to outsiders because of legal restrictions (e.g., to avoid anticompetitive collusion among companies), safety and security concerns, or trade secrets that need to be protected. And of course, opening internal markets to external participants requires care. For example, opening talent markets may require different strategies than simple salary bands for deemphasizing price. But in other areas, it's likely that internal markets will be accessible more swiftly outside the firm's border.

There's a certain irony in the fact that firms hope to compete against the data-rich market by incorporating some of its key features. And, ironically, too, as firms incorporate elements of data-rich markets to save the firm, they enable the capture of huge streams of data, which in turn could also fuel a renewed attempt toward decision automation.

FIRMS LOOKING TO RESPOND TO THE RENAISSANCE OF the market can choose between two strategies—option one, to automate decision-making, and option two, to rearrange a firm's organizational (and thus decision-making) structure. Amazon, for example, could opt to incorporate data-rich markets into its organizational structure—a strategy well-aligned with the marketplace that Amazon is for many of its customers. But Amazon could also choose automation of decision-making, given its data-driven organizational and decision-making setup. The ultimate choice of strategy will likely be driven by where Amazon sees its core competitive advantages located. If Bezos perceives Amazon's hierarchical organization as a key advantage, moving away from it may not be enticing. And as Amazon's artificial intelligence systems and massive computing resources are cutting edge, so the Everything Store may be well positioned to push toward automating managerial decision-making. Getting to automated decision-making, however, will take some time. In the interim, Amazon may want to experiment with organizational change.

What's important to understand is not so much the concrete strategy that a firm should pursue. It's critical that the right strategy depends less on a firm's lofty vision than on its capabilities at any point in time—and how they can be translated into competitive advantages. It's a challenging choice that all firms will face. Moreover, the two strategies for firms we outlined aren't mutually exclusive. Companies may adopt strategies that combine some elements of both. This will surely lead to important improvements, but it likely is not enough to save the firm as we know it, because long term, both options undermine the traditional idea of the firm as the preeminent and most efficient mechanism of *human* coordination.

Yes, option two emphasizes restructuring the organization, a thoroughly human construct. But decision-making power is shifted away from managers to an internal market system in which decisions depend on a powerful flow of rich, multidimensional data paired with machine learning systems and in which transactions are enabled by powerful algorithms. The process may still require human participation, but over time humans may not be making many of the decisions. Option one leaves organizational structures in place: no market encroaches on the firm, but through automation, managerial decision-making increasingly shifts from humans to machines.

In the very long run, therefore, firms may model themselves on two archetypes: one in which a firm owns most of the resources needed for its operation, and still may employ humans, but is managed and run mostly by machines, and another in which firms rely on market mechanisms but in the process shed most of their organizational functions. This latter type of firm may eventually end up as an *organization of one*—a single person who coordinates the market mechanisms and thus becomes nothing but a market participant. Neither organization will employ many humans to coordinate its activity, at least not compared to today's firms.

In the meantime, existing firms as well as new start-ups have their work cut out for them: decide what decisions to delegate to machines and harness the power of the market to improve the way they coordinate.

– 7 –

CAPITAL DECLINE

W HEN METEOROLOGIST BOB CASE LOOKED AT SATEL-lite imagery and weather data for the northern Atlantic Ocean on October 27, 1991, he was taken aback. Having worked for the National Oceanic and Atmospheric Administration for decades, he quickly realized the danger: a huge cold front, moving north, was about to hit a high-pressure system, moving south from the Canadian coast. "These circumstances alone could have created a strong storm," Case said. "But then, like throwing gasoline on a fire, a dying Hurricane Grace delivered immeasurable tropical energy to create the perfect storm."

And a perfect storm it was, bringing hurricane-force winds and waves up to one hundred feet high. It caused more than $200 million in damages. By Case's reckoning, such an event would hit the New England coast only once every fifty to a hundred years.

Case's work forecasting this perfect storm was immortalized in a best-selling book and blockbuster movie, and it anchored the term in our vernacular.

Today, large swaths of the banking sector are themselves facing a perfect storm. Like the confluence of the three distinct phenomena that led to the 1991 weather event, three distinct but reinforcing threats may turn banking on its head. Each one alone is a challenge, but taken together, they may wipe out a significant portion of the industry.

The first is the structural weakness of the banking sector, which was exposed by the subprime mortgage crisis beginning in 2007. This banking crisis was caused at least in part by information that was either incorrect and incomplete or wrongly interpreted. According to one estimate, more than $8 trillion was lost, and there were extensive bank bailouts in a number of advanced economies. In the United States, the Emergency Economic Stabilization Act of 2008 led to the federal government's earmarking more than $700 billion in loans to help ailing banks. The capital infusion quickly stabilized the US banking system at the height of the crisis

In the United Kingdom, Germany, and Italy, governments recapitalized banks with funds in the hundreds of billions of dollars but also essentially nationalized banks by buying shares of the banks rather than purchasing nonperforming securities that the banks held. Banks suffered from the subprime mortgage crisis more than any other economic institution, and the ensuing loss of confidence in their stability put the sector under great pressure to adapt.

The second threat was born of the Great Recession, which commenced in 2008. Central banks in many nations responded to economic woes by lowering interest rates, and as rates hit rock bottom, banks in some countries even charged negative interest for deposits. Savers were not the only ones affected. In practice interest rates on deposits cannot be lowered much beyond zero (negative rates are highly unpopular and may lead to withdrawal of deposits). These low interest rates also reduce the interest spread—the difference between the rate banks pay their creditors and the rate they charge their customers—and thus banks' margins.

Banks were already suffering from the increased competition that has shrunk banks' margins in the United States from almost 5 percent (in 1994) to around 3 percent (in 2016). The situation is at least as bad in Europe, where the interest spread has dropped to 1.4 percent. At the same time, new banking regulations have increased banks' overhead costs. Many commercial banks have not been making high profits anymore from traditional credit and savings operations. In fact, researchers at the German central bank predict that with the interest spread staying low, only one in five banks in Germany will earn a decent return on cost of capital in the coming years.

Profits are also negatively affected by changes on the payment services side. Online banks employ an order of magnitude fewer employees per customer account than traditional banks, resulting in a drastic cost differential. As a result, they can offer payment services at much lower fees to their customers. In addition, formerly lucrative international money-transfer services are under competitive threat, thanks to start-ups such as TransferWise. More generally, due to a costly information infrastructure stemming from the twentieth century, banks can no longer compete effectively with digital rivals such as PayPal and Apple Pay, which offer their services to consumers at very low cost. Their overhead is small: PayPal has no expensive branch network to pay for, and Apple Pay can piggyback its service on proprietary security technology built into more than a billion iPhones.

Banks have responded with comprehensive cost cutting, continued automation, and the shrinking of their physical footprint. In 1990, commercial banking employed just under a million and a half people in the United States, but the sector never recovered from the financial crisis of 2007 and today no longer employs as many people as it did in the early 1990s. The situation is worse in Europe. As of 2016, 27,000 fewer bank branches employed 212,000 fewer people than before the Great Depression. In 2015 alone, one in ten private banks in Switzerland vanished, including centuries-old brands such as Bank La Roche (founded in 1787). The sector's crisis is affecting well-known names, too: Germany's Commerzbank will cut one in five positions by 2020, and Italy's large UniCredit bank will close 26 percent of its branches.

These trends alone would suggest that banking is in dire straits. But as if they aren't bad enough, a third element looms on the horizon. It may topple even some of the supertankers in banking, and it has to do with the role of money.

IN DATA-RICH MARKETS, PARTICIPANTS NO LONGER USE price as the primary conveyor of information. Of course money still stores value, and participants still use it to pay. But if money will no longer be necessary as an efficient information shorthand, one of the central functions that money has performed in the economy will be gone. Its role will further diminish as the transformation to data-rich markets continues—as standard data ontologies, matching algorithms, and machine learning

systems advance and as market participants embrace more efficient transactions based on a richer flow of information.

For most of us, there won't be an immediate difference in our
daily activities. We'll still pay with money, although the lowered
cost of banking will get us more bang for the buck. But the real
difference will be felt on markets. When we are better able to
compare what potential transaction partners have to offer along
many dimensions, we'll change how we weigh information. This
will lead to vastly more efficient market transactions, as price
will become only one data point among many, rather than a bell
buoy in an ocean of noise.

The ramifications of this shift are profound for everyone in
the business of money. As market participants look for and consider richer data, fewer of them will depend on money, and they
won't be willing to pay as much for its informational function
as in the past. This hurts the entire financial services sector, although the pain, as we'll explain, isn't evenly spread.

Because money is no longer that important for greasing
the wheels of the markets, our view of the economy will evolve.
Rather than equating markets with money, and the economy
with finance capitalism, in which money rules supremely, markets will be understood to surge because of rich data flows (not
money). Finance capitalism will be as old-fashioned as Flower
Power. Some may miss it dearly, but that fondness will be little
more than nostalgia.

Richer information requires new ways to convey it and communicate with it. Consider the simple case of a store window.
In the past, a display window mainly contained the products on
offer and the price tags associated with them. In the future, we'll
expect to learn more about each product (and, likely, the seller).
Because this information can't be expressed by a simple number

on a single piece of paper, it must be conveyed some other way—digitally (for sure), and quite likely wirelessly, then analyzed by an app that allows us to search for the best possible matches, given our preferences.

The infrastructure necessary to convey such detailed information is being built right now. Many advanced digital market platforms already offer an impressive flood of multidimensional information, while physical markets, including brick-and-mortar stores, are still pondering ways to adapt the technology for their needs. Retailers, for example, have their hopes pinned on what's called augmented reality, which enriches what we can see on the sales floor by providing additional information about the available goods. It's like a much-improved version of Google Glass and will highlight perhaps the three products in a shop that best fit your preferences, and you will learn about them by looking around. For our purposes, it's not as important to predict exactly which technical solution will offer us the richest information and in what form as it is to realize that the solution won't depend on the established infrastructure of money and price that banks and other financial institutions have built.

The shift away from price signifies a monumental change: separating the act of payment from the provision of (much) information. Money-based markets afforded money an outstanding role in the economy because it was central to all stages of a market transaction, from the search for and identification of a potential partner all the way to the completion of the transaction.

Banks and other financial intermediaries acted as the servants and facilitators of a system that was so fundamental and crucial to the market that money became almost synonymous with the market system. And it began to spread through the entire sector. The finance industry enjoyed being perceived as both a holder of wealth and a fount of insight. That wasn't wrong. Banks were the conduits to a lot of price information flowing through the market. To the extent that banks analyzed that information, they were even able to assist their customers in decision-making. It did not produce perfect results—far from it, in fact. But because of the access to information that banks enjoyed, trusting their recommendations was often better than deciding in ignorance—at least until the next financial crisis.

Better is the enemy of good. Using rich information streams to make decisions is superior to relying on money alone. As the economy shifts to data-rich and thus more efficient markets, most of the information necessary for these markets to function will no longer flow through banks. Banks will still handle the completion of transactions through the transfer and storage of value and perhaps even contribute modestly to the overall flow of information. But the informational center of gravity in markets is moving away from money—and thus away from banks.

As payment service providers, banks will offer essentially a commodity service and will be forced to compete against savvy new entrants that are not saddled with legacy infrastructure. It's like letting a lot of air out of your inflatable life raft during a storm: you may still stay afloat, but with such an impaired craft, it's much harder to get anywhere, let alone to your intended destination.

Perhaps the storm may be calmed a bit by the thicket of regulations with which players in the financial services sector must comply. Such compliance is not only costly, it also insulates these players somewhat from competition, while new entrants struggle with regulatory currents. And yet, perhaps at first counterintuitively, regulation may also speed up the reshaping of the banking sector. Although much existing banking regulation is complex and difficult to deal with for digital start-ups, regulators are beginning to understand the information dimension of banking and to appreciate the power banks currently possess, thanks to their role as information conduits. Correctly, they assume that such information might translate into transactional inefficiencies, and there is a push underway for measures to counter this. In the European Union, for instance, the Second Payment Service Directive, coming into force in 2018, mandates that any bank must, if a customer requests it, provide competitors as well as third parties with the customer's digital data that banks hold. The goal is to make it easier for bank customers to switch banks (much like mandated number portability with mobile phones eased the pain of switching providers), as well as to create a new market of financial information intermediaries. Having access to rich banking data, they can then assist consumers in their decision-making. It's yet another example of a disentangling of traditional banking (which is getting more commodified), and value generation based on rich and comprehensive information. Intriguingly, at least the European regulators seem to think that such innovation in financial information will be led by new entrants rather than traditional banks.

Other traditional financial intermediaries, especially those focused further on the informational role of money and banking, will fare even worse than banks in this new environment. They have no payment function to fall back on, no life raft to save them. Think of the stand-alone broker and the conventional insurance agent, for instance; they may drown in the sea of information that rewired markets unleash.

THE DEMISE OF MONEY AS THE MARKET'S PREEMINENT conveyor of information will also prompt a decline in the role of capital. In our market system, financial capital is key because it is such a fungible factor of production: when necessary, it can easily be exchanged for a much-needed resource and vice versa, thereby enabling efficient resource use.

But capital also conveys information. It signals to the world that a company has an asset at its disposal that it can exchange for other factors of production. It connotes freedom of choice as well as relative power. Outside investment increases a company's flexibility, and it also conveys further information—about the prowess of the company as well as the trust that an investor has in it.

Sometimes, the informational dimension of an outside investment may be more valuable than the capital inflow itself. When a highly respected venture capital (VC) firm such as Sequoia Capital invests in a Silicon Valley start-up, it's akin to conferring a peerage in nineteenth-century England: the recipient gets immediate name recognition and often gains additional market value as a result.

As markets embrace diverse information streams, these two functions of capital—information and value—are no longer necessarily intertwined. Rather, they will more frequently be separated. The point here is not to suggest that in the future, capital will have no role to play. Capital in its function as value will continue to be useful in our economy. But it will no longer be the only information game in town.

When we disconnect the two functions of capital, we realize that their relative importance depends on context. For example, at times a company may need an actual influx of capital. But in other situations, giving a signal to the market that a widely trusted expert believes in the business is precisely what is needed. To be sure, not all signals are equally honest. Talk is cheap and a recommendation nobody believes is not worth much, especially when compared to an endorsement backed up by a much-coveted fat check. In contrast, an honest signal carries a cost that deters potential abusers. Money isn't always an honest signal, nor is it the only one. When money is abundant, for instance, the informational value of such a signal is greatly reduced. And as MIT professor Sandy Pentland has argued, a wide variety of signals, including those generated from network and social media data, can be honest.

The rewired market will have no problem ingesting and conveying such signals, and market participants will have no difficulty factoring such signals into their computer-assisted decision-making (of course, this does not guarantee perfect choices, only that choices reflect all available preference information). This is particularly important when money is abundant and investment opportunities are limited. At such times, a capital investment itself is no longer as strong a signal of endorsement; it has lost a portion of its informational value. Consider, for instance, a VC firm in times of abundant availability of capital that is unable to invest in a start-up it likes, because the investment round is oversubscribed. The firm may then invest elsewhere, but that investment isn't a signal that the company it is investing in is the best choice, only that it is the one that was available.

A world of abundant capital may sound unreal, but there has been renewed talk recently about such a situation developing for venture capital activities around the world, as the volume of deals has increased, reaching record levels not seen since the dot-com bubble of 2000. In general, more money is becoming available for capital investments as investors are trying to identify opportunities that offer higher returns than the rock-bottom interest rates available through conventional and conservative investment mechanisms. Attracting capital, especially for start-ups in the right location, is far easier today than it was in earlier decades. One CEO of a Silicon Valley start-up commented that the company, absent an urgent need, raised money simply because it could. At the same time, fewer investment options exist on conventional stock markets. In the United States, the number of listed companies is down from more than 9,000 in the late 1990s to fewer than 6,000 in 2016.

If capital is abundant, but fewer companies are looking for capital, supply outstrips demand on capital markets, and this means returns on investment are plummeting. This spells the end of finance capitalism as we know it—the mental association of working markets with huge investor returns. And it's unlikely the good days of finance capital will ever come back: as markets turn data rich, there is less need to signal with money. The economy will thrive, but finance capital will not thrive with it on account of the shift from money-based markets to data-rich ones. With the market economy advancing with the help of data, we may no longer label the future "capitalist" in the sense of power concentrated by the holders of money. Ironically perhaps, as data-driven markets devalue the role of money, they prove Karl Marx wrong, not Adam Smith.

In essence, data takes over from money, and investors will be paying the bill. This means trouble for investors in general, but in particular for those who saved and invested for a steady monthly income in retirement. They banked on money retaining its preeminent role and now face an unexpected financial short-fall. This may lead to the widespread dissatisfaction among those who diligently saved every month of their working lives; they will feel cheated out of their dream of a cozy retirement. It's a challenging predicament for policy makers, too, because as we know from previous disruptions caused by earlier innovations, there is no alternative to the painful adjustment that lies ahead. There is no obvious policy solution, either, that preserves the central tenet of financial planning that we should be saving for retirement throughout our working lives.

If there is reassurance to be had, it is that although data-rich markets will cause a drastic shock to the system, with thousands of billions of dollars in individual holdings evaporating as rates of return drop and investments lose their value, this shock will likely be one-time, rather than recurring. Once capital has been devalued and our expectations of the anticipated returns from it are reset, capital's value will likely hold steady, rather than continue to slide. This is no solace, of course, for current investors, especially those a decade or so away from retirement and thus fully dependent on solid capital returns. These investors may end up as the generation hit hardest by finance capitalism's out-sized promises and abrupt demise.

In the long run, however, data-rich markets will help investors to better identify opportunities that match their preferences and are less clouded by human bias. New intermediaries will rush in to fill the demand, using sophisticated matching tools and machine learning systems to analyze a flood of information and translate it into fact-based advice. We'll still need financial advice, but it will likely come from a machine rather than a human being. Because it's essentially software running on rich data, the digital investment adviser is able to work with us on our personal computing devices (including smartphones), rather than in an office. That creates the opportunity for an unprecedented level of privacy if we so desire: we can grant a machine learning system access to our very personal data (including our investment behavior) so that it can distill our investment preferences and find optimal matches without the need to share it with anyone else.

The inverse is possible, too. We can permit a system to use our data for other purposes such as training itself, or making general market predictions, in return for a reduced fee. Operating a machine learning system is also cheaper than relying on a human adviser, and at least in principle, it can be set up to ensure that no hidden fees exist that distract from optimal advice (such as when an adviser gets paid a percentage of each transaction, which creates an incentive to suggest unnecessary transactions). And, because such systems can be designed to fit any requirements, existing service bundles (think of an investment adviser who also executes the transaction) are more likely to be broken up. They will make room for an ecosystem of investment guidance with different providers offering different services that can be easily combined even by individual investors. Finally, as all services run on comprehensive data, it's natural for them to make available data about themselves, so that investors can choose the most appropriate intermediary to work with. The market for investment advice, much like the future market for investments, will become thoroughly data rich. Investors may lament the passing of the old days of high returns, but they will have much to gain from data-rich markets that offer improved matching.

FOR BANKS AND CONVENTIONAL FINANCIAL INTERMEDI-
aries, on the other hand, the reduced role of money poses a
complex challenge. So far, they have reacted by pursuing two
main strategies: the first focuses on cost cutting, primarily
through automation; the second on reinventing themselves as
information intermediaries in data-rich markets.

Lowering costs begins with a shift from a physical infrastructure to a more digital one. As more customers bank online
and on mobile devices, banks no longer need a comprehensive
branch network or as many bank clerks and tellers. Actions to
reduce the cost per transaction, whether for investment management, lending, or payment follow. As discount brokerage
Charles Schwab demonstrated in the 1970s and 1980s, if transaction costs are sufficiently low, money can still be made even
with reduced fees.

But in the twenty-first century, banks aren't competing against
Charles Schwab–type firms. Instead they face a new generation
of start-ups that are relentlessly utilizing digital technology to
extract insight from data and provide services at rock-bottom
prices. By applying high-frequency trading technology to conventional stock markets, Silicon Valley–based Robinhood Markets offers its more than 1 million customers the opportunity
to trade stocks on US exchanges with zero commission. This is
possible because the actual cost of an electronic trade today is
so low. By forgoing any expensive physical infrastructure (like
storefronts or a large support department) Robinhood Markets
can depend on interest generated from deposited but not yet invested funds for its income. Even after lowering their costs, it
will be hard for banks to compete against free services.

A similar development is under way for payment solutions.
Here, established digital players such as PayPal, Apple Pay, and

WeChat in China, aided by mobile-payment start-ups such as Stripe and Square, are having the banks for lunch. In addition, they get all the valuable transactional data, before passing on the minimum data necessary to banks for the actual transfer. In the payment business, we are seeing a repeat of the evolution of mobile phones: telecom operators used to see all traffic data from their customers but did little with it. Today, they're merely conduits that can't (or are not allowed) to peek into traffic anymore, and all the value generated from the data traffic is captured by others.

Some fintech start-ups, such as Coconut in the United Kingdom or Holvi in Finland, focus not on lower fees but on innovative additional services. They are targeting niche markets (small businesses for Holvi, freelancers in Coconut's case) with a service bundle of payment and bank accounts that's highly customized. For example, Coconut offers customers the ability, whenever receiving or making a payment, to swiftly (re)calculate taxes and to put money aside to pay them. Holvi's services include free integrated invoicing and bookkeeping.

Some banks have sought to push cost cutting and automation much further by teaming up with or investing in companies that are working on alternative payment systems. In financial circles, Bitcoin (as well as blockchain, its underlying technology) has not only caused fear but also instilled hope that banking can be saved—although it's unclear how. Banks advocating deeply decentralizing technologies for transferring and holding value such as blockchain may not yet fully appreciate that these technologies obviate the need for the centralized service they are offering.

Overall, cost cutting may sound smart, but in banking it is as constrained by organizational setup and internal structures as it is for any other firm, and banks are already beginning to realize that. Lowering costs may help the banking sector in the short term, but over the long term, it may amount to little more than rearranging deck chairs on the *Titanic*.

Responding to the cost of money disregards the demise of its informational dimension. As markets turn data rich, money is no longer needed to facilitate most of the information flow. No digital currency is capable of fundamentally altering that, not even the most advanced blockchain technology. Essentially, these are solutions for a different problem.

But even as a medium of exchange, money may no longer hold an absolute monopoly. If markets teem with information that facilitates transactions, that information itself holds value. Every time it gets used, it creates insight and greases the market. Such market information turns into a valuable resource that is useful not only for one specific market participant, but for the market as a whole.

As long as its application is wide and its value high enough for a sufficiently large number of market participants to want it, and as long as the cost of exchange is sufficiently low, in the future we may see transactions paid in data rather than money. In a way, we are already doing that every day: when we use Google's search engine or log on to Facebook, we tolerate advertisements that serve as payment for the personal data collected about us. In fact, Google and Facebook would not be what they are without the billions of users who pay for their services with personal data. Similarly, in an increasing number of cases, companies contract with outside services to have their data analyzed for them, and they pay these services essentially with data, by letting them reuse the data for other purposes.

This isn't the end of money. Data has an important drawback as a medium of exchange: much like satchels of salt and gold coins and much unlike paper money, data is valuable in itself, complicating its role greasing market exchanges. Therefore, money will continue to play an important role facilitating exchange (and central banks will need to continue to manage its supply).

These changes in the role that money performs won't happen overnight, but they demonstrate that the non-informational functions of money aren't necessarily insulated from any disruption, either. If overall the importance of money is declining, strategies that focus mainly on cost cutting while continuing to rely on money as the fundamental lubricant in markets may work in the short and medium term but will be of less utility long term.

A GROWING NUMBER OF BANKS AND OTHER FINANCIAL intermediaries are employing a very different strategy, aiming to reinvent themselves as data-savvy intermediaries. They are even teaming up with new entrants in the financial services sector, in preparation for a world after money.

There is some logic in the fact that banks are providing capital for fintechs, companies that use data technologies to provide financial services, many of which aim to push conventional banks off their pedestals. The banks' bet is: if you get disrupted, you should at least own some of the players that take away your business. In 2015 alone, these fintechs attracted investments exceeding $19 billion worldwide. Some pundits have described the frantic activity as a fintech bubble. Although a number of fintechs focus on payment solutions, many of them focus squarely on offering disruptive innovation in two areas we'll examine briefly: lending and investment planning.

For decades, money lending has progressed from a matter of personal trust—think of a community bank manager deciding whether somebody gets a mortgage or not—to a decision driven by a single statistic: an individual's credit score. Reducing trust to a number may seem to make it easier for banks to choose their borrowers. But as we know, the reality is more sobering. Conventional credit scores are very heavily based on past credit transactions and constrained by a paucity of data. When few data points matter, errors in that data get amplified, and a credit score could grossly overstate or understate a problem. It is a ridiculously crude way of judging a person's ability and willingness to repay a loan.

To remedy this, a whole ecosystem of new loan providers has sprung up that not only ingests, but also provides, a lot of information. For example, SoFi, a fintech start-up, which originally focused on student loans, factors many data points into its prediction of creditworthiness, allowing it to offer low interest rates to individuals with limited credit history; Kabbage offers a similar service to small businesses. Shifting from conventional credit scores to a risk model that analyzes more numerous and diverse data points is like moving from money-based markets to data-rich markets: in both cases, we abandon the idea of condensing complexities and instead use technology and automation to guide decisions based on comprehensive and rich data sets from a wide variety of sources. This translates into a better assessment of actual default risk, and thus enables SoFi to offer loans to many of its customers at lower rates than what conventional lenders charge.

SoFi's model has been a resounding success: by 2017, the lender had funded more than $16 billion in loans, saving its customers, it says, an estimated $1.45 billion in interest. Another new entrant, Upstart, is a fintech that uses educational data in addition to traditional credit scores to assess credit risk and capture the risky end of the credit market. And fintechs such as Avant and ZestFinance (founded by former Google CIO Douglas Merrill) combat payday loan-sharking. By using machine learning and analyzing a huge number of data points per credit applicant, they believe they can calculate risk far better than traditional banks and thus can offer loans to individuals who otherwise would be at the mercy of the payday loan industry. In 2016, Chinese Internet giant Baidu invested in ZestFinance, with the aim of bringing data-rich consumer credit scoring to China.

Fintech start-ups also disrupt traditional investment management. For example, Stash has been pushing to break apart the share as the smallest possible unit of investment, instead enabling their customers to buy fractions of a share. That way, consumers can spend a small amount of money according to a specific investment strategy. It's a bit like unbundling individual song titles from albums.

Many fintech start-ups suggest that they have far superior preference-matching and preference-extraction tools than conventional financial advisers do. Betterment, for example, touts its ability to identify capital losses, thus enabling its customers to lower their tax bills. Another contender, SigFig, gathers and analyzes data about the investments its users make through brokerages and identifies alternative funds with similar risk profiles, giving its users more choices. It also calculates how much investors pay in fees to those brokers (and how much they could save by switching).

Fintechs have given birth to a whole new niche sector of data-rich platforms, such as ZuluTrade and eToro that offer customers a way to select from and copy the investment activities of many thousands of other traders. The idea is that these platforms offer a way for investors to learn from and emulate traders with matching preferences. These platforms make money by taking a cut from transactions they stimulate. Other sites, such as PeepTrade, offer customers access ("peeps") to decision information of successful traders and take a cut from each trade that "follows" a successful trader's strategy.

Financing and investment come together through fintechs offering peer-to-peer lending. On these platforms consumers lend to consumers (or in the case of platforms like Funding Circle, to small businesses), and the platform facilitates these lending matches, although often the matching process is still relatively unsophisticated. Zopa, one of the pioneers in peer-to-peer lending, has successfully helped arrange financing for $2 billion, but the idea has really taken off in China, because of the country's bureaucratic traditional banking system. Thousands of platforms have opened, and the most successful one, Lufax, has already eclipsed Zopa in total financing volume. The total market for peer-to-peer lending in China is estimated to have surpassed $100 billion in 2016. Kickstarter and its competitors, such as Indiegogo, offer a related service. Kickstarter alone has helped start-ups generate direct sales in excess of $3 billion, with one in three projects being successfully funded (and only about 15 percent of funded projects eventually failing). Recently, Kickstarter has teamed up with equity crowdfunding platform MicroVentures to offer backers a chance to buy equity in small businesses. What's interesting is that Kickstarter built a platform for start-ups to go beyond a simple purchasing or funding transaction and offer comprehensive, rich, and continuous information to backers, quite a bit like an early data-rich market, so as to provide them with a lot of information in their decision-making and also keep them in the loop later.

Much like firms in general, fintechs approach digital technology and data abundance differently. Some, like those offering low-cost money transfers, offer inexpensive versions of existing services; they essentially bet on the past and hope to take advantage of the financial sector's perfect storm by better braving the waves. Others are squarely focusing on rich data streams. Betterment and SigFig, as well as a number of peer-to-peer lenders, fall into this category: they pair comprehensive data with user preferences and algorithms to identify the optimal transaction partner. Over the long term, fintechs in this class—and perhaps the banks teaming up with them—aim to position themselves as information intermediaries that have capabilities beyond those of money and price.

As they shift the focus from money to rich data, these businesses not only undermine the belief in the power of money and banking; many fintechs also employ *markets* for tasks that once were the purview of traditional large banks. This only further underscores the general shift in the economy that we have asserted throughout this book—from firms to markets.

So far, the results from banks joining forces with fintechs have been mixed. In part this is surely because we are still at the beginning of the shift away from money, so there is a lot of uncertainty, a lot of trial and error, before stable and successful business models emerge. It's a bit like e-commerce in the mid-1990s, before the dot-com boom. But the challenge may also be deeper and more cultural. Although in theory, banks should be very comfortable working with lots of data because they've gathered a wide variety of financial data for many decades and operate large data stores containing detailed customer information, they haven't done much with the data they have. In this context they are rich in data but poor in insight.

This may be a remnant of earlier times, when analyzing data was difficult and costly. And banks are traditional. Their ethos generally is to preserve rather than to risk, and their staffs reflect that ethos. Perhaps it also reflects a strong customer preference; maybe people don't trust their banks enough to let them use the rich personal data banks have collected to create new products or services. But it is likely structural as well: if an organization has been focused on helping money lubricate the economy, it will have a hard time thinking far beyond what has made it successful. As much as the single-minded focus on money turned banks and related financial intermediaries into preeminent institutions of finance capitalism, so, too, does it constrain imagination and prevent the embrace of a data-rich future.

A counter example suggests it does not have to be this way. In the early days of investment banking, more than a century ago, these banks were very small partnerships helping companies to find outside investment. They also helped mostly very large private or institutional clients to identify the most appropriate targets in which to invest. They were matchmakers, and their success rested on privileged access to information. They had long-term relations with customers, and an investment bank's success was linked to the tacit knowledge and comprehensive network of its partners. The need to maintain their good reputation kept many of them honest and the valuable information they had access to confidential. They were essentially data-rich information intermediaries in an analog age.

Over time, and accelerating with the 1960s, the sector changed, in part because large banks began to compete with traditional investment banks. These new entrants lacked the privileged information networks, but they had scale and a much tighter focus on money. A number of investment banks responded by reinventing themselves as institutions of money, rather than information: they did this by abandoning partnerships and issuing listed shares, by merging and growing drastically in scale, and by dramatically increasing leverage. They turned themselves into highly leveraged banking institutions. When the subprime mortgage crisis hit in 2007, three of the world's largest investment banks—Bear Stearns, Lehman Brothers, and Merrill Lynch—collapsed, and many merged with conventional banks, utterly remaking most of the sector.

A few investment banks, however, had resisted that shift toward money. They remained information intermediaries and did well. More followed, and today a growing number of highly specialized boutique firms utilize the latest in digital technologies, working with big data analytics firms such as Contix and Kensho and utilizing machine learning systems to do what investment banking originally did: offer information rich with insight about optimal investment transactions to market participants.

This case may reflect the opportunities as well as the perils of reinventing oneself. It may dampen our optimism that the banks and other core institutions of finance capitalism that are so focused on money can turn themselves into innovative startups. But we may understand this case also as a metaphor for the rise and eventual fall of money-based markets, and the success of information intermediaries over monetary ones. We suggest this may happen throughout the financial services sector as finance capitalism is replaced by data capitalism. Venture capitalist Albert Wenger, whose firm has funded many successful start-ups in the financial sector from Kickstarter to SigFig, likens the fate of traditional banks in the age of rich data to another image of a tempest threatening a ship—a "Spanish Galleon full of raided gold sinking in a storm." It has access to all the capital but lacks the insight, based on information, to circumnavigate the perilous weather.

– 8 –

FEEDBACK EFFECTS

THE AIRBUS 330 ROSE MAJESTICALLY INTO THE EVE-ning air on June 1, 2009, as it lifted off from Rio de Janeiro's international airport. The 216 passengers on board Air France Flight 447 looked forward to an uneventful journey to Paris.

Commercial passenger flights have achieved an amazing safety record, thanks in no small part to powerful computers and well-trained cockpit crews. Together they form an elaborate feedback system. The flight computers process huge volumes of data from dozens of sensors, keeping the plane on track (itself a feedback loop) and flying safely, while the pilots monitor the computers, examining rich data presented to them about the plane's position, trajectory, and health. Both check each other— the computers can ignore pilot commands that would endanger the aircraft, and the pilots can switch off computers' flight control if they need to. But neither happens often. In fact, computers have gotten so good at controlling planes (and adjusting to compensate for erroneous human commands) that aircraft manufacturers build them into all commercial airliners. As a result, the computers fly the airplane, mostly, and the pilots keep their watch.

It was that way for the first couple of hours of Flight 447. Then, in pitch-black darkness as the plane approached thunderstorms over the Atlantic Ocean, airspeed sensors iced over and stopped functioning. The computers realized the problem and partially disengaged in order to let the pilots take over. The pilots could now fly the airplane as they needed to, without the computers second-guessing them. For unclear reasons, the copilot decided to climb the aircraft by rising its nose, although the plane had already gotten close to its maximum altitude. That slowed the plane down dangerously; it risked stalling, losing lift, and falling to the ground.

The computers sounded an alarm—a voice announced "Stall" to the pilots, but the pilots, fatally, failed to react. As the nose of the plane continued to rise farther and farther, the stall alarm stopped, because the computers, no longer trusting the extreme position data they were receiving, had decided that something was wrong with the data (it wasn't). As the pilots were trying to understand what was going on, they entered a fatal feedback loop: whenever they lowered the plane's nose, the computers, now thinking the data about the plane's raised nose was plausible again, sounded the stall alarm; whenever the pilots then raised the nose, the stall warning stopped—not because the plane wasn't stalling, but because it was stalling too much for the computers to trust. In a way, both sides—humans and computers—acted perfectly reasonably. The machine warned when it considered the data to be reliable, and the humans reacted to the stall warning. They never realized that the aircraft was plummeting toward the sea. Moments later, all on board had perished.

Until the end, the pilots were trying to figure out what was going on. In most cases, the feedback loop of such a human-machine system works well, and even if it doesn't, it fails gracefully. But systems based on complex feedback loops are tricky: they work so well in so many routine cases that we are tempted to disregard—even forget about—any built-in risk of extreme failure.

Seven decades earlier, MIT professor Norbert Wiener, a child prodigy turned accomplished mathematician, conceived the general theory of feedback and its role in helping humans and machines control their actions. Feedback loops lie at the very core of Wiener's concept: collecting and interpreting feedback data enables control over a system and adjustment of its goals. Wiener's big idea was that any system can be steered in the direction we want as long as enough feedback loops are built in. Conceptually, it was a huge leap to understand how machines can work independently—or, to put it in today's words, run autonomously. It laid the groundwork for technical developments from the guidance systems of intercontinental nuclear missiles (and the Apollo moon lander) all the way to modern adaptive machine learning systems. But Wiener also looked at and worried about catastrophic failures of feedback systems that could, as the story of Air France Flight 447 highlights, be triggered by unexpected situations, or if elements of a feedback system were caught in an erroneous loop.

Wiener's concept *of* control in systems has also fostered the desire *for* control: if something can be controlled, it ought to be, and often in a centralized fashion. The mathematician had anticipated this in choosing to name the study of systems control "cybernetics." Its Greek origin, *kybernete*, means "governor." And after World War II, Wiener turned into one of the earliest critics of the cybernetics revolution he helped to unleash. In his book *The Human Use of Human Beings*, Wiener examines information flows as key enablers of feedback-driven cybernetics and openly worries about adaptive systems not "because of any danger that it may achieve autonomous control over humanity," but because they "may be used by a human being or a block of human beings to increase their control over the rest of the human race."

Data-rich markets resting on feedback-driven systems exhibit similar qualities. They work well in most cases but can become shockingly dangerous either because of a lack of diversity when learning, or from the veiled centralization of control. It is these two dangers that we must protect against through appropriate governance measures.

MARKETS ARE CHAOTIC AND UNPLANNED. THEIR ESSEN-tial quality is that they are built on a myriad of individual decisions. There is no central control, but like any social mechanism, markets aren't empty containers. Every feature of a marketplace, whether physical or social, whether internally adhered to or externally enforced, shapes which transactions can take place. The famous Champagne fairs of the Middle Ages required merchants to follow a set of norms or be expelled and lose the opportunity to transact.

Market behavior is also molded by physical design. Covering part of a market square with a roof held up by pillars allows merchants to easily enter and leave the trading space, and the impressive cloth halls that once dotted major European cities made it easier for buyers and sellers to brave the elements. This meant the markets could be operated well into the colder seasons of the year, which shaped what would be sold, when, and by whom.

It's no secret that markets with the right features work better than others. And though many norms that shape markets depend on specific circumstances, there are a few key principles that transcend specifics. Perhaps the most important of these principles is the diffused nature of decision-making. If this principle is violated, a market loses its ability to coordinate human activity efficiently.

Unfortunately, in practice many markets have gotten concentrated over time with negative effects. In the steel industry of the late nineteenth century, a small number of large producers colluded to dictate prices. A similar situation arises when many sellers face a single buyer: think of many small farmers having to sell their milk to just one local dairy. The dairy effectively gets to set the price it pays for milk. Concentrated markets are problematic because they deprive many participants of better deals, while excess profits accrue to the most powerful players.

Online markets seem particularly vulnerable to concentration. Just consider that about four out of five search requests from desktop computers and nine out of ten requests from mobile devices go to Google. Think of Amazon's share of more than 40 percent of online retail revenues in the United States, and Facebook's almost 2 billion users worldwide. And these are only the very big names. In smaller niches, too, there are high market concentrations. GoDaddy is the largest domain-name registrar on the Internet, around four times as big as its largest competitor. WordPress dominates blogging, while Netflix rules movie streaming.

At least since the days of Karl Marx, there has been a vigorous debate regarding the exact reasons for this dynamic. We suggest that three distinct effects are often at play when markets become concentrated: scale effects and network effects, both of which have been well researched, and a third effect, fueled by the role of feedback for adaptive systems that we therefore term the "feedback effect." Each of these effects is an outgrowth of market participants' strategies to expand their profits.

During the Industrial Revolution, manufacturers realized the potential of producing in volume. If a factory churned out 1,000 Model T cars a week rather than one hundred, the cost per car came down as fixed costs of production were distributed among a larger number of units. What began in manufacturing spread to many other sectors of the economy, including retail and services. Supermarkets, fast-food franchises, and retail chains mushroomed in the second half of the twentieth century because they all aim to lower their costs by increasing their volume. That strategy has worked, and consumers have reaped the benefits of the scale effect: lower prices and a wider range of products.

Network effects can be seen in the telecommunications industry. By 1890, the phone had taken markets in the United States by storm. It provided such a swift and simple way to coordinate with others that no large business could do without it. But because there were several rival phone companies, each with its own proprietary network, managers often had multiple phones on their desks so they could call business partners who were on different networks. When the phone market consolidated around AT&T at the turn of the century, customers realized that every new subscriber who joined the network increased the system's utility for everyone already on it because it added to the number of people they could potentially reach. It provided a strong incentive for even more people to subscribe to AT&T. It was as if the service got better as people were added to it—even though the service itself did not improve, only the opportunities to use it. Today, this network effect (economists sometimes prefer the term "network externality") is very familiar to us. It's what allowed the Internet to dominate the flow of digital information and what has driven the success of social media platforms from Facebook to WeChat to Twitter and Instagram. The network effect also enhances the value of market platforms—from eBay to Alibaba, to ride-hailing companies such as Uber and Didi Chuxing, and from Tinder to peer-lending pioneer Funding Circle—although the exact value a new participant adds depends not only on that person but also on the existing players in the market. For example, when a man joins an online dating platform for heterosexuals whose membership is 90 percent men, his usefulness is minimal. Far more valuable would be new female members—at least for the men on the network. Similarly, in a market with an abundance of sellers, every new buyer will be particularly welcome.

The third effect, though related to scale and network effects, occurs whenever computer systems use feedback data to learn. When we react as Google autocorrects a spelling, our response creates feedback information that improves Google's spell checker. IBM's Watson gets better at recognizing skin cancer the more skin cancers it "sees." The most popular products and services improve the most because they are fed the most data. In such a context, innovation is no longer about breakthrough ideas but rather about collecting the greatest amount of feedback data.

The scale effect lowers cost, the network effect expands utility, and the feedback effect improves the product. Each lead to significant benefits for market participants: they can lower the cost of production, grow the value of their services, or offer a good that continuously evolves seemingly by itself.

These effects are not mutually exclusive, either. Companies can realize two or even all three of them at the same time. Consider Amazon: because of its sheer scale, it can fulfill customer orders at low cost. Network effects make Amazon a thick market, with lots of buyers and sellers, and many customers who leave valuable product reviews for others. Each additional customer adds value to the community. Finally, Amazon uses adaptive systems and feedback data to hone its recommendation engine, as well as its intelligent personal assistant, Alexa. Apple's iPhone is another case in point. Because it can mass produce the phone, Apple can keep profit margins high while still holding to a price point that's acceptable to consumers. A growing number of iPhone users have led to a vibrant app market. And Siri (among other services) continually improves, thanks to a huge and increasing volume of feedback data. Combined, these three effects have led to great advances in the kinds of products and services available on the market. Unfortunately, though, they have also been driving concentration—the deadly poison for market efficiency.

Market concentration is on the rise, especially in the United States and the United Kingdom, and is becoming more prevalent in continental Europe in a wide variety of sectors and industries. Big companies are getting bigger, while (despite our seeming infatuation with entrepreneurship and start-ups) business dynamism, driven by innovative disruptors challenging entrenched incumbents, is ebbing, especially in the high-tech sectors.

For more than a century at least, nations have put in place comprehensive rules to guard against the dangers of concentration. But these rules do not prohibit concentration in general. Antitrust and competition experts understand market concentration to be deeply suspect but not by itself a reason to intervene. The concentration caused by scale and network effects has been tolerated as long as large players did not abuse their market power. That's why the antitrust lawsuit against Microsoft in the United States, filed in 1998, which almost resulted in the breakup of the company, focused on Microsoft's *behavior* rather than just its position in the market. Similarly, the recent antitrust case against Google in Europe, which claimed that its search engine results list Google's services above those of its competitors, targeted the company's behavior rather than just its market share.

In addition to scrutinizing corporate behavior, regulators also look at how difficult it is for a new entrant to compete against large, established players. If it's not very hard, no intervention is needed, because new competitors can enter the fray. If entry is very difficult, regulatory action may be in order.

The scale effect once presented a huge hurdle for new entrants in markets like manufacturing or chain retailing, in which scale matters. In these markets, achieving scale traditionally required prohibitive initial investments. But with the rise of venture capital and low interest rates, raising funds has gotten easier, enabling start-ups to grow swiftly in both scale and scope. Moreover, thanks to the plummeting cost of information processing and storage, especially through cloud computing, the initial investment necessary for start-ups is often much lower than it was in the industrial age.

In contrast, network effects remain problematic. Even with a lot of money, start-ups often have great difficulty attracting customers. The only path to success seems to be through innovation:

to offer something substantially better than what the incumbent is offering. There's a robust debate among lawyers and economists as to the extent that innovation offsets the network effect. Some point to the persistence of dominating platforms, such as Microsoft Windows for PC operating systems and Facebook for social media. Others highlight the fact that Facebook unseated the previous incumbent, MySpace, and is now being threatened by Snapchat, a clever start-up based on the innovative idea of vanishing messages. They also point to the rise of Linux against Windows and the fact that PC operating systems are less important today, as computing is often done on mobile devices and tablets (areas in which Microsoft does not enjoy a dominating market share).

So far, regulators have focused mostly on the scale and network effects and have yet to grasp the gravity of the threat to markets posed by the feedback effect. Services based on machine learning systems fueled by feedback data "buy" innovation at diminishing cost as the user base grows. It feels strangely alchemistic: turning a by-product of usage into the raw material of improvement, like converting lead into gold.

This has huge implications for market competition. Incumbents that command substantial streams of feedback data thanks to their large customer bases have a powerful source of machine-based innovation at their disposal. Start-ups cannot hope to compete with them effectively, as they lack the volume of feedback data that can drive product development.

A growing number of experts, among them legal scholars Ariel Ezrachi and Maurice Stucke, are worried that machine learning systems are undermining competition. They advocate for measures that go beyond regulating anticompetitive behavior. Others have even suggested that large companies should be forced to "open up" their algorithms and let competitors as well as the general public have access to them, much as the code for open-source software is made available.

The call for open algorithms misjudges the root of the problem and would do little to counter market concentration. Algorithms are both the recipes for learning and its outcome. As recipes, often they are already widely accessible and in the public domain. As concrete outcomes of data fed into adaptive systems, they may change each time a system learns from new data, so having access to them is but getting a condensed glimpse into the past. It's like price in money-based markets: it holds some information but lacks detail. Algorithms alone aren't enough to enable small competitors and new entrants to compete against incumbents because algorithms aren't the raw material they need for their adaptive data-driven systems to learn.

The mistake doesn't lie in the idea of sharing but in what is being shared. Rather than algorithmic transparency, regulators wanting to ensure competitive markets should mandate the sharing of data. To this end, economists Jens Prüfer and Christoph Schottmüller offer an intriguing idea. They suggest that large players using feedback data must share such data (stripped of obvious personal identifiers, and stringently ensuring that privacy is not being unduly compromised) with their competitors. Calculating the effect of such mandated data sharing over a wide spectrum of scenarios, they see an overall net benefit in most cases, especially when one incumbent is close to dominating a market.

Building on this idea, we suggest what we term a *progressive data-sharing mandate*. It would kick in once a company's market share reaches an initial threshold—say, 10 percent. It would then have to share a randomly chosen portion of its feedback data with every other player in the same market that requests it. How much data it must make available would depend on the market share captured by the company. The closer a company is to domination, the more data it would have to share with its competitors. This is different from the sharing mandates in banking that we mentioned in Chapter 7 (even though the ultimate goal of ensuring competitive markets is the same). The point there was to lower switching costs; here it is to spread data as the source of innovation.

Large companies won't lose the benefits of the feedback data they collect; their products will still improve as they gather more data. But by having to share a portion with others, the value derived from the data gets spread around. This benefits smaller competitors and helps them compete against large players. Moreover, by implementing a progressive lever, data sharing increases whenever concentration increases. It's a feedback mechanism to counter the feedback effect: the more concentration threatens competition, the more vigorously the data-sharing mandate kicks in.

Although large players could also request access to feedback data from smaller companies, the biggies would benefit much less from additional feedback data relative to their smaller brethren. And the mandate to share feedback data with every player in the market avoids creating incentives for players to misstate or exaggerate their market share. Consider a market with two large competitors: one has 45 percent market share and the other has 40 percent market share. The remaining 15 percent is scattered among many small companies. If the data-sharing mandate only went one way—from the biggest company to the smaller companies—only one of the big players would have to share with the other, creating an incentive to game market share. But a universal sharing mandate would benefit both large players because each would have access to the data of the other—even though, in relative terms, the companies benefiting the most would, appropriately, be their small competitors.

Our plan for progressive data sharing is targeted at the growing number of companies that use feedback data and adaptive machine learning to improve their service offerings, from Google, Facebook, Apple, and Microsoft all the way to Tesla. At first this may look like a small slice of the overall economy. Yet product improvements based on data-driven adaptive systems are so impressive that an ever-increasing number of companies will adopt them, thereby expanding the purview of progressive data sharing.

The dangers of concentration persist and arguably even increase as we move from money-based to data-rich markets. This requires continued vigilance against illicit behavior on the part of powerful players. It also necessitates new measures independent of behavior, such as this progressive data-sharing mandate, to counteract the feedback effect—the network-effect equivalent of the data age.

IN DATA-RICH MARKETS MACHINE LEARNING SYSTEMS combined with matching algorithms will assist us in decision-making (or in some cases even decide on our behalf). But they aren't without structural deficiencies of their own.

Markets suffer not only when they are concentrated, but also when many market participants make the same flawed decision—for instance, when many humans, susceptible to the same bias, make the same mistake. While market decisions are decentralized, the decision makers are still human, suffering from similar cognitive constraints. As we suggested in Chapter 4, these learning systems will—to the extent we want them—help us overcome some of the biases that plague us. This will lead to data-rich markets being less vulnerable to systemic erroneous decisions, and resulting breakdowns. That's a significant improvement over traditional money-based markets.

But it hinges on machine learning systems not only working well, but working sufficiently *independently* of each other. It's a crucially important requirement. Consider what could happen if adaptive data-driven machine learning systems for assisting us with transaction decisions were all supplied by one or a handful of providers. They would be able to shape market decisions more comprehensively, but also less transparently than any market concentration we have ever experienced in the past. If the Achilles' heel of data-rich markets is the danger of central control of adaptive systems helping us decide, we must adamantly guard against their control by one or a small number of firms, lest we be forced to accept a commercial Big Brother.

Our vigilance should also go beyond the issue of control. Even if the companies supplying us with such decision-assistance systems are perfectly benign, a single point of failure embedded in the structure of data-rich markets would make them (and us) uniquely susceptible to outside attacks. It's as though everyone were driving only one kind of car: What do we do when we discover that someone has tampered with the brake system? In the context of Air France Flight 447, all flight computers in modern Airbus airplanes exhibited the same behavior, and thus after the terrible accident, all pilots flying Airbus aircraft had to be trained to understand correctly what the stall warning was telling them and when. Homogeneity of the systems we employ amplifies their flaws and can lead to a systemic vulnerability.

To avoid such a potentially catastrophic systemwide fault, participants in data-rich markets must be able to make meaningful choices from among a wide variety of decision-assisting systems, designed and maintained by a variety of providers. If each is designed and developed independently of the others, it is less likely they will share common flaws. But achieving this heterogeneity of decision-assistance is a challenge. Markets for systems fueled by feedback data tend to concentrate, wiping out the kind of robust competition we need in order to guard against systemic defects. Therefore, ensuring competition among decision-assistance systems despite the market's inherent concentration tendencies is key to the sustainability of data-rich markets.

A progressive data-sharing mandate, the solution to market concentration among machine learning systems in general, is *the* mechanism of choice: when feedback data from large players is available to smaller competitors, then innovation in decision-assistance systems is not concentrated at the top, and smaller players may remain in business.

To summarize: Like other markets, data-driven ones require rules (and their stringent enforcement) to make sure that decision-making remains decentralized and markets remain efficient. Unlike traditional money-based markets, data-rich markets may be less susceptible to human systemic biases and thus to devastating breakdowns. But this decision-making advantage can only be sustained if the systems used to assist market participants are varied and diversified, offering humans real choice. This requires unique regulatory measures directed at the root cause of the problem: the highly uneven distribution of feedback data.

DATA-RICHNESS IS INFORMING HOW MARKET PARTICI-
pants report on themselves to others. Because reality is varied
and multifaceted, so, too, the instruments we use to capture it
must be sensitive and capable of communicating nuance. This
means that reporting and accounting practices as well as rules
concerning transparency must become more comprehensive and
detailed.

Accounting standards, for example, have long mandated that
the value of certain assets on a company's balance sheet should
equal their historical cost. That was easy and straightforward: if
land was bought for $1 million, its book value would also be $1
million. But that value did not necessarily reflect reality: the land
might have gotten much more or less valuable in the meantime.
The book value as a single figure provided information about
the historical transaction but did not convey much about its
current value. Figures in balance sheets could therefore not be
trusted—not because they were wrong but because they might
be outdated.

As part of an accounting reform, beginning in the 1990s in
the United States, and followed in many other nations, certain
assets must be priced at "fair value," which is often the same as
their current market price. This results in a balance sheet that
more accurately reflects a firm's position at a specific moment in
time. The challenge is that market prices fluctuate. As a result,
not only are balance sheets that "mark to market" (as this accounting practice is called) outdated the day they are published,
the changes may also indicate temporary fluctuations that don't
accurately reflect the financial health of the company. Land value,
for example, may drop because of an increase in car traffic next
to it, but that could be temporary, caused by a detour during the
construction of a new highway. If those changes are captured in

the balance sheet of the company that owns the land, it might show a steep drop followed by a huge increase, and the swing might cause investors to sell or buy shares in the company or banks to call its loans, even when the firm's financial situation had not materially changed. Critics of fair-value accounting even maintain that it has contributed to the severity of global financial crises, including the subprime mortgage crisis in the United States in 2007–2009.

Whereas fair value may be a more useful figure than historical cost, it, too, captures value in a single figure, as a snapshot in time. It may be easy to read, but because conceptually it is rooted in informational oversimplification, it is hard to act on. With the rise of data-rich markets, therefore, we must develop new accounting practices that convey much richer, more detailed information, such as how long a firm anticipates owning a certain asset, the volatility of asset prices, and the relative risks associated with them. This will give third parties a better sense of a company's real value, capturing not just a moment *in* time but also painting a picture *across* time.

The need for a more comprehensive view is not limited to questions of value and price. Balance sheets could reflect much more than just the financial status of a company. They could also include numbers representing its energy use, environmental impact, and labor standards. The availability of such comprehensive data about companies would make it easier for investors in data-rich markets to find suitable matches that reflect their preferences beyond the typical fundamentals. And what works for investors would also work for transaction partners in general. The prerequisite is that the relevant data must be made available—and in a standardized form. That is where rules about reporting and transparency kick in.

SO FAR WE HAVE HIGHLIGHTED GOVERNMENT'S ROLE IN facilitating the flow of information and balancing information deficits by mandating that more information becomes accessible, through sharing mandates and reporting rules. In some situations, however, regulators have been tempted to *reduce* the flow of information so that one side doesn't have more than the other and can't translate that information imbalance into favorable transaction terms. This is what conventional information privacy legislation does. The goal is admirable, but the practical trouble with such collection-limitation rules is that they create incentives for companies to circumvent them or get consumers to consent to indiscriminate collection (as many of us do, when in the process of signing up for an online service we click "okay" without reading the fine print).

In practice, collection-limitation measures rarely redress information imbalances and thus would also fail to facilitate competitive and effective data-rich markets. Instead of limiting collection, therefore, government's task in data-rich markets may be to limit the ways in which information from one side can be *used* by the other—in other words, to focus on the limits of data *usage* rather than collection. That way, using data to improve optimal matching could be encouraged, but using data to prompt inefficient information imbalances could be discouraged. Within the information-privacy community, a vibrant debate about such a shift in focus from collection to use is already under way. Data-rich markets may underscore the need to consider such a shift, and to deliberate on details.

Government agencies will be responsible for implementing the elements of the efficiency-enhancing regulatory framework we have outlined in this chapter. Whether an existing agency is asked to expand its jurisdiction or a new agency must be created depends on the national context and is less important than the agency's effectiveness of enforcement. Whatever institution takes on these duties and enforces the rules in data-rich markets needs to have not only the organizational capacity, the staffing, and the required investigative and enforcement authority but also the appropriate expertise. That will neither be easy nor cheap. Talent is scarce, and regulators will have to compete with the fat paychecks and prestigious work environments offered by firms and start-ups from Wall Street to Silicon Valley. But there's really no alternative. Government needs professionals with data-analytics expertise, the "quants," if it does not want to risk market failure. We do not call for a bureaucratic expansion lightly. But without organizational enforcement, data-rich markets will be vulnerable to a dangerous concentration of decision-making power and control.

At least as threatening for liberal democracies is the temptation for government to use the mechanisms of data-rich markets, such as decision-assistance systems, to take over control of the economy. The world had a hint of what might be possible about half a century ago, when mainframe computers still ruled supreme.

SEPTEMBER 11, 1973, MARKED THE END OF ONE OF THE most ambitious governmental experiments in history. It commenced in July 1971, when Fernando Flores, then head of the Chilean Production Development Corporation, asked British operations research and cybernetics pioneer Stafford Beer to build a computer system that would assist the Chilean government in planning its economy.

A year earlier, physician Salvador Allende had become the first elected Marxist president in Latin America. He had been aiming for a socialist "third way" for the economy, eschewing both free markets and a Soviet-style command economy. His platform featured land reform and the nationalization of large-scale industry as its main components. But the newly nationalized industries required management.

This was the reason Flores sought out Beer. They were an odd pair: the socialist economic administrator (who later would become Chile's economic and finance minister), desiring to centrally manage most of Chile's industry, and the iconoclastic British business-school-professor-turned-management-consultant, who had a penchant for cigars, chocolate, and whiskey. But they shared a dream: a new form of governance fueled by accurate data and swift feedback, both for those who direct and for those who are being directed. It was a vision of coherent nationwide decision-making based on a combination of organizational structures and modern technology. By conceiving of people as a collective in need of government support rather than as individuals choosing their own paths, it resonated with Flores's idea of a socialist utopia, and by employing the latest communication and control mechanisms at a grand scale, it appealed to Beer's fondness for cybernetics.

The plan they hatched was called Cybersyn (nicknamed Synco in Spanish), a socio-technical system designed to govern and direct Chile's industry. The idea was that every day, four hundred nationalized factories around the country would send data to Cybersyn's nerve center in Santiago, the capital, where it would then be fed into a mainframe computer, scrutinized, and compared against forecasts. Divergences would be flagged and brought to the attention of factory directors, then to government decision makers sitting in a futuristic operations room. From there the officials would send directives back to the factories. Cybersyn was quite sophisticated for its time, employing a network approach to capturing and calculating economic activity and using Bayesian statistical models. Most important, it relied on feedback that would loop back into the decision-making processes.

The system never became fully operational. Its communications network was in place and was used in the fall of 1972 to keep the country running when striking transportation workers blocked goods from entering Santiago. The computer-analysis part of Cybersyn was mostly completed, too, but its results were often unreliable and slow. In part this was a structural problem, because information flowing to Cybersyn was neither comprehensive nor accurate. But it was also a technology problem, because the processing power available at the time was grossly insufficient, leading Cybersyn to flag problems late—sometimes days after decisions should have been made.

Cybersyn died a premature death when the Chilean military staged its coup against President Allende and destroyed the system's infrastructure. The underlying idea, however, has lived on, and with it the hope that the technology can be used to govern an entire nation.

Cybersyn's central governance mechanism isn't the kind of central control that Stalin wielded and that led to the Great Famine of 1932–1933, which caused some 7 million deaths. Stalin wanted widespread industrialization, and Soviet economic authorities directed everyone to make it happen, irrespective of individual needs and wants. The resulting economy of scarcity might have delivered satisfaction to a few elites but resulted in misery for the masses. These structural inefficiencies have discredited central planning as a general economic policy, and almost all nations have abandoned the idea. Of course, in national security and public safety, perhaps education and health care, central planning has proven to work. But these are exceptions to the rule—islands of central planning in what in many countries are seas of decentralized coordination through markets.

In contrast, the temptation offered by a project like Cybersyn—in equal parts enticing and abhorrent—is to use feedback-driven systems based on the latest digital technology to achieve government control, not by issuing brutal and unilateral central directives but by using machine learning systems in data-rich markets to subtly shape individual decision-making.

At first, this looks relatively benign. As we have explained, data-driven machine learning systems can help us reduce human bias to the extent that we want them to. But wouldn't it be better if these systems could eliminate human biases regardless of what an individual desires? And why restrict a system's corrective activity to human biases? Couldn't we design such systems to inject additional values such as civility, justice, and equity into the decision-making process? Recently, some experts have argued for subtle "nudges" to coax us to transact appropriately. With widespread machine learning systems running on rich data, we could turbocharge nudging into a highly individualized (and thus precise) process of shaping people's perception. Think of Fox News joining forces with Facebook. People would still render their decisions on markets, but because they would all be advised by a single system, society could progress toward common goals along a coherent path.

This is a modern version of Cybersyn, but with a sinister spin. At least Cybersyn was transparent: the centralization of planning and decision-making was obvious to all Chileans. By contrast, government control of adaptive machine learning systems in data-rich markets retains the trappings of decentralized coordination and the appearance of free will, but turns Norbert Wiener's powerful concept of cybernetics into Big Brother riding data-rich feedback loops. It's precisely what Wiener was worried about. The system, even if perhaps appearing to promote liberal values, would make George Orwell blush and the East German Stasi salivate: seeming freedom on the outside but total state control on the inside.

This is the dystopia that our society faces as we transition to data-rich markets. It is why we have to become much better at applying traditional antitrust legislation in the digital economy. But it is also why new measures such as our proposed progressive data-sharing mandate are so crucial in protecting the decentralized nature of decision-making, so we can preserve not just markets but an open society in general.

– 9 –

UNBUNDLING WORK

W ALT MARTIN WAS IN THE MOOD TO CRACK A JOKE. "I've got to practice my yoga," he told a *Wired* magazine reporter who had joined him in his eighteen-wheeler as it cruised at 55 mph on I-25 in Colorado in October 2016. Just like that, the truck driver scrambled into the sleeper cabin behind the truck's cab to stretch out and check his tablet computer while the truck and its 50,000 cans of Budweiser kept heading south toward Colorado Springs. The journalist, who was there to cover the first autonomous truck delivery in the history of logistics, noted: "The drive was as mundane as the beer in the trailer."

Just after he guided the truck onto the highway, Martin had pushed a button labeled "Engage" and a $30,000 system built by San Francisco start-up Otto—with LIDAR laser detection, radars, and high-precision cameras, military-precision GPS, and a powerful, data-crunching computer—took control of the vehicle. Otto, which had been acquired by Uber in the summer of 2016 for $680 million, offers so-called Level 4 autonomous driving on highways. With it, there is no need for a human driver to stand ready to take over, at least not for technical reasons. The company describes its "vision" as turning a conventional truck into a "virtual train on a software rail." So far, the system works very well, except for unforeseen events (like sudden heavy thunderstorms) that it has left to future versions of the software to solve.

As Walt Martin was fishing for his tablet on this trial delivery, Otto's vision seemed achievable. As long as the weather cooperates, Martin needs only to drive the few miles from AB InBev's brewery at Fort Collins to the interstate and a few miles more from the highway's exit to the delivery ramp at his destination.

The successful test drive demonstrates that the technology for autonomous driving on highways is almost ready to launch. Because highways are where truckers delivering 70 percent of all freight in the United States spend almost all their time, improving road safety through autonomous driving could spare the lives of some of the 4,000 people killed every year in the United States by trucks. But increasing economic efficiency— in terms of lower fuel consumption, improved utilization of expensive equipment, and reduced labor costs—is the real motivation behind the adoption of autonomous driving systems. No wonder that firms around the world are vying for a piece of the action, from start-ups such as Otto and Embark to truck manufacturers such as Daimler's Freightliner, Volkswagen's Scania and MAN, and Volvo, which is partnering with Otto.

At the same time, Otto's cofounder, Lior Ron, found it necessary to reassure America's truck drivers that the system won't take away their jobs because they will be "needed in terms of supervising the vehicle." Otto says they want to make a trucker's work less stressful, but not redundant. This may be what the many millions of truck drivers in the world want to hear, but given the cutthroat competition in logistics, can it be true?

In the United States, truck driving is one of the few remaining jobs that doesn't require a college degree but provides, according to Bureau of Labor Statistics data, a relatively decent median annual income of more than $40,000. It comes with more job security than most jobs do, because a trucker in Minnesota can't be replaced by a factory worker in Shenzhen. And truck driving has gotten fairly sophisticated: trucks now come equipped with power steering and braking, automatic transmission, satellite navigation, and modern fleet management systems. Today, the trucker is less a manual laborer than a controller of sophisticated digital equipment, even though the decision-making is mostly routine. Truckers also have to handle a fair amount of administrative work.

As odd as the comparison might be to Japanese insurer Fukoku Mutual, which uses IBM's Watson to replace claims assessors, what truckers do isn't all that different from the type of middle-income desk jobs that we anticipate will disappear as automation becomes more widespread. And even if new job opportunities open up for truckers, they'll require different skills.

Data-rich markets, and data-driven machine learning systems in general, are bringing vast changes to the human workforce and further fuel a reconfiguration of the labor market that is already under way. In the United States, the percentage of people participating in the labor force has declined from its peak in 2000 to a rate that was last seen in the 1970s. The decline coincides with the rise of digital technologies, the Internet economy, and the early adoption of data-rich markets. Studies, some of them quite speculative, forecast depressing employment figures for white-collar workers for the next decade in all advanced economies and many others. Of course, that's not new. With every increase in automation over the course of the nineteenth and twentieth centuries, many millions of jobs have been eliminated. Simultaneously, because of human ingenuity, even more jobs have been created. As manufacturing became more automated, the services sector grew.

The question today is whether this will happen again. With a well-developed services sector that itself may face the challenge of increasing automation, what is there to employ the middle-class workers displaced in data-rich markets? Is this the advent of a "second machine age"—the neat phrase to describe the coming displacement through automation of white-collar jobs coined by MIT professors Erik Brynjolfsson and Andrew McAfee? It is likely, as we suggested in Chapter 6, that there will be less work for humans in the future; but no matter what happens to overall labor force participation, it is almost certain that the types of jobs available will be quite different from the jobs people hold today.

The situation is actually even more dramatic when we look not just at the number of people participating in the workforce but also at the amount of the nation's income allocated to worker and employee compensation. In the United States, this "labor share" has declined considerably since the 1980s, from 67 percent of value added to 47 percent, much of that drop happening after 2000. The United States is not alone. In most advanced economies, labor share has declined. It has also fallen in the large economies of India and China. In fact, economists estimate that labor share globally has been dropping since the 1980s.

This trend has puzzled economists. For decades, labor share had stayed relatively stable. In a competitive economy, there is a simple reason that this should be the case. When the number of people employed goes down but the number of goods produced stays the same, productivity per employee goes up: each worker produces more value. In due course, this should translate into an increased wage, as companies vie for more productive employees. Higher labor costs, by definition, increase the labor share and cancel out the effect of the productivity-related displacement of workers. However, since the 1980s, labor share did not rebound but continued to decline. At the same time, in the United States, capital share—the part of a country's overall income going to capital—seemed to increase. As a result, less of the US economy's total income is going to workers and more seems to accrue to those who provide capital: investors and bankers.

Sophisticated analysis of the labor-share decline has shown that the drop isn't caused by one labor-intensive industry declining while another less labor-intensive industry is growing. Labor share has dropped in almost all industries, pointing to something that affects them all: digital and data-processing technologies. And because labor share is not declining evenly across all income levels—the share of the highest-earning 1 percent of workers is falling more slowly than the rest—automation appears to be the major driver of change. It displaces blue-collar and low- and middle-income white-collar workers first.

Although the direct effects of technical advances, especially digital technologies, are crucial, indirect effects play a role as well. For example, labor-participation statistics capture workers employed by a firm as well as the self-employed, thus obfuscating fluctuations between these two categories. The number of self-employed workers with no employees has increased substantially in the United States, from 15 million in 1997 to nearly 24 million in 2014. Researchers at the Brookings Institution believe that in part, this shift is the result of a growing "gig economy." When self-employed white-collar workers take on temporary gigs with limited or no benefits, assignments that are often coordinated through digital platforms, they have little bargaining power over their incomes. Unlike unionized workers in old-fashioned manufacturing, they are rarely organized, and the supply of workers often outstrips demand. This contributes—albeit to a limited extent—to overall wages being out of sync with the actual value labor generates.

We are still in the early days of the data age. The rate at which workers are made redundant by data-driven automation is bound to accelerate in the coming years. If the link to labor share is real, this will prompt a further reduction in the share of income accruing to workers, while it enriches investors and banks.

Concerns about the shrinking role of labor, and the shift in income distribution, have already caused alarm in many corners, from economists such as Thomas Piketty, whose 2014 book, *Capital in the Twenty-First Century*, a stinging critique of capitalism, became a global best seller, to populist movements (such as those led by Marine Le Pen and Donald Trump) that promise to eradicate the plight of the displaced worker. And two sets of relatively conventional ideas, one distributive and the other participatory, are being advanced by policy makers and debated in many nations as a response to this troubling trend.

As income seems to shift from labor to capital, the proposed measures on the distributive side aim at taxing the sources of this automation-driven income. This can be achieved, for instance, by taxing machines such as computers running machine learning systems (or more precisely taxing the value that machines add) that make this new wave of automation possible. Colloquially called a "robo tax", this mechanism would supplant payroll taxes as workers are being replaced by machines. The goal is not necessarily to tax the economy more, but differently. In early 2017, the world's richest person, Bill Gates, announced his support for a robo tax. At the same time, the European Parliament discussed and ultimately rejected a proposal to apply a robo tax throughout the continent. Although it was promoted as a way to fund social programs designed to soften the effect of white-collar automation on the European workforce, it was dismissed because of fears it might stifle innovation.

A somewhat different distributive strategy is to impose a tax not on the value generated by machines but rather on the capital that supposedly claims a rising share of the economy's income. This could be achieved, for example, through an increase in the capital gains tax or by instituting a new wealth tax. The latter is the more popular option and the one that is being advocated in several European nations. In most proposals, these distributive measures would keep overall tax revenues flat or increase them only incrementally, just enough to offset the decrease in tax revenues resulting from a decline in labor share.

These distributive measures seek to establish a more just tax system that focuses on where the value add is generated, and to offset some of the negative tax consequences of this upheaval in labor markets. But in practice, their effects may be limited. Even robo-tax proponent Bill Gates focuses on maintaining funding for social security programs as the overall workforce contracts. He doesn't claim that the robo tax is a solution to the social problems this contraction will bring about. Similarly, a wealth tax may feel fair, but experts predict it will yield only a modest volume of tax revenue and do little to remedy social disruption.

Participatory responses, by contrast, propose a set of policies that support the retraining of workers who have been made redundant by decision-assistance systems in particular and by the digital transformation more generally. Their proponents suggest that it is possible to increase the labor share by making sure that any demand for workers with special skills (including, for example, machine-learning experts, but also specialized human care workers for a graying population) can be met. The assumption is that labor share decreases not only because of automation but also because jobs demanding specialized technical skills cannot be filled with the expertise available in the current labor market. But this reliance on retraining workers assumes that the data revolution will unfold much as the Industrial Revolution did: initially displacing many workers but eventually leading to a variety of new jobs. Retraining accepts that job creation happens when human ingenuity is paired with the magic of the market.

It has a positive, empowering ring, suggesting that people can rejoin the workforce if they acquire new skills. But as our economy transitions to data-rich markets, automation will increasingly replace white-collar workers, like the insurance assessors at Fukoku, who have substantial expertise and those with administrative and managerial skills.

It may be relatively easy to look at the skills that are needed today and design training so people can acquire them, but that strategy probably will not prepare workers for what will be needed tomorrow. And training for skills that today we may perceive as useful in the future runs the risk of anticipating needs incorrectly. Who knows exactly what new jobs will emerge (and thus what skills will be needed) in the age of data-rich markets? The dynamism of change in the currency of skills means that what may be much in demand in three years may no longer be so sought-after in a decade. Young people joining the workforce now will likely have to retrain their skills multiple times during their work life, and to do so at an accelerating pace. Without sufficient insight into exactly what the economy will need, retraining runs the risk of conveying the wrong skills and thus producing the wrong supply for the demand. It's a bit of a crapshoot: one might be right, but much more likely, one will be off the mark.

THESE DISTRIBUTIVE AND PARTICIPATORY MEASURES are relatively conventional. All are adaptations of policies that already exist in many advanced economies around the world. They are not without merit but do come with drawbacks. There is a far more radical alternative measure being put forward, in the form of universal basic income. UBI, as it is affectionately called by its proponents, has garnered surprising support, particularly among leading figures in the high-tech sector. "Superangel" investor Marc Andreessen, the coauthor of Mosaic, one of the first widely used Web browsers, is in favor of it. And so are New York–based Albert Wenger, another highly successful venture capitalist; start-up incubator impresario Sam Altman; and Elon Musk, the brash but congenial cofounder of PayPal and CEO of Tesla. Silicon Valley isn't alone in its enthusiasm for UBI, but it is Silicon Valley's digital and data-driven innovations that have given rise to the idea.

There are innumerable variations, but the core idea is similar. Everyone receives a monthly check for a fixed amount that would be sufficient to pay for food, clothing, basic education, a warm, dry home, perhaps even some form of health insurance. People made redundant in data-rich markets would no longer have to worry about how they and their families could get by. In this sense, UBI has both a distributive and a participatory dimension to it. Because money is being taken from taxpayers (there is a lively debate which tax resource would be most suitable to fund UBI) and given to everyone, it is distributive. And it is participatory, because its goal is not only to provide people with some of their basic needs but to enable them to rejoin the workforce at less than full time.

The idea of a universal basic income has been circulating among economists and progressive politicians ever since the late eighteenth century, when Thomas Paine proposed a basic income for everyone above fifty. In the middle of the twentieth century, radical pro-market economist and Nobel laureate Milton Friedman suggested a negative income tax that had many of the distributive qualities of a universal basic income but would have been somewhat more complex to administer. In 1972, Democratic presidential candidate George McGovern openly advocated for a universal basic income. He was attacked by incumbent president Richard Nixon and ultimately had to withdraw his plan, but Nixon then proposed his own family-assistance program that would have been close to being a UBI for a large segment of society if it had not died in the Senate.

Universal basic income is an idea that is attractive to both liberals and libertarians. The former see in the UBI a comprehensive welfare program that gives everyone sufficient funds for a life of dignity and relieves those who struggle in poverty. Unlike existing needs-based social programs, which require people to apply for them, a UBI would conceivably carry no stigma, because every adult would receive the same amount.

Libertarians, on the other hand, embrace UBI because for them it kills two birds with one stone. First, it eliminates the need for a huge government bureaucracy to determine individual needs and calculate individual payouts. As everyone gets the same UBI, there is no need for any bureaucratic case-by-case assessment. In many proposals, UBI would be a direct, simple replacement of existing needs-based welfare programs. It would empower people to choose for themselves what they want to spend their monthly incomes on rather than give them specific amounts earmarked for purposes (such as housing subsidies). And, second, by giving the same amount to everyone, a UBI benefits even the very rich (not that they need it), somewhat sidestepping the rhetoric of income redistribution, which has been at the core of progressive social policies for more than a century. In addition, by giving everyone just enough to get by and no more, a UBI would not necessarily eliminate the incentive to work.

Limited experiments in administering a UBI are underway in Finland and the Netherlands, and they will produce some empirical data about the effects of a UBI on human motivation. Switzerland held a national referendum on a UBI, but voters rejected the very generous scheme that was proposed (around $2,000 per month for every Swiss citizen). Beginning in 2016, start-up accelerator Y Combinator even sponsored a small project in the United States designed to investigate whether receiving a basic income would have an effect on people's desire to work. Canada experimented with a version of basic income back in the 1970s, when it gave a monthly check to every eligible family in the small town of Dauphin, Manitoba. That led to improvements (albeit modest) in education rates and reductions in hospitalization and teenage pregnancy. There wasn't much of an impact on the number of people participating in the workforce because nobody was quitting their job to live off the government paycheck. The experiment did not result in sufficient data, however, of the effect on those already unemployed and seeking work.

Critics have pointed out that many UBI proposals are highly regressive: they replace welfare and thus eliminate programs that provide support to people with special needs, from children with learning disabilities to the physically challenged. After all, a single wheelchair—let alone a year's worth of chemotherapy—is a lot more expensive than what one could afford on UBI alone. By getting rid of needs-based programs in return for UBI, the poor would essentially subsidize the middle class and the affluent, because those in need would get less than they presently do, whereas the more affluent would still get a monthly UBI check.

The biggest challenge of a UBI, however, is finding the necessary funding for it. A modest universal basic income of, for example, $12,000 per year for every adult in the United States would cost around $3 trillion, or more than twice the total 2016 federal budget for Social Security and roughly 10 percent of gross domestic product. Even eliminating all Social Security payments in lieu of a UBI wouldn't be sufficient. Of course, this may not deter some proponents of a UBI, who point to additional budgetary savings such as a diminished bureaucracy, the overall social and societal benefits, and the resulting economic stimulus to justify spending such a huge sum on a single government program. But it puts a weighty practical damper on the notion that a UBI can easily be financed.

UBI is not necessarily a bad idea. But more important than taking sides in this debate is understanding the significance of the UBI movement as well as its limitations. Perhaps it doesn't come as a surprise that high-tech innovators are looking for creative ways to respond to the social challenges posed by data-driven markets and the reconfiguration of our economy. Ingrained thinking is something not even Silicon Valley can escape: if you believe in disruptive innovation, it may become your standard response to any problem.

But it may not be all that innovative an idea. If data enables us to go beyond money, why does the social innovation designed to solve the problems caused by data-driven markets emphasize money? Why are we reintroducing, through a UBI, a simple, fixed monetary solution in situations that quite obviously require an assessment of needs beyond money? After all, the whole idea of data-driven markets is to leave behind the straitjacket of money and the condensation of information into single price points and shift toward a form of human coordination that offers superior matching of our preferences. It is odd that those who otherwise are champions of rich and comprehensive data reduce the discussion to a conventional single monetary dimension. In that sense, a UBI would be less regressive than retrograde, gazing backward rather than looking forward. Of course, we understand that people still need money to pay for food and shelter. The puzzle isn't that UBI offers a basic income to people but that it doesn't offer *anything beyond money*. That seems to artificially limit the scope of what can and ought to be done.

THESE THREE SETS OF POLICY MEASURES—CONVENTIONAL, distributive, and participatory ones, as well as the more radical UBI—are based on certain economic assumptions: not only that labor share continues to decline, but that labor share will decrease *while* capital share increases, and a dangerous imbalance in our economy exists. Moreover, these policies assume that there is a direct inverse relationship between the two: if one decreases, the other increases. Otherwise, it would make no sense to tax capital (either machines or wealth) in lieu of labor. As it turns out, these assumptions are far less certain than previously thought.

Intrigued by the inverse relationship between labor share and capital share, experts have looked more closely at the supposed rise of capital share. Northwestern University economist Matthew Rognlie has shown that much of the increase in capital share evaporates when one excludes the housing sector. Another economist, Simcha Barkai, has pointed out that conventional calculations of capital share assume fixed rates of return, while in reality, over many years, interest rates have continued to decline, reaching a very low level. Barkai recalculates capital share based on actual interest rates and finds that capital share actually *decreased* substantially since 1997—around three times as much as the decrease in labor share. This throws a gigantic monkey wrench into the widespread belief that the data age is destroying labor while profiting capital. So what's going on?

There may be several factors at work. For one, the cost of data processing has plummeted as processing power, storage capacity, and network bandwidth have all greatly increased since the late 1990s. As a result, data processing requires less capital, leading to a reduction of capital share. The low cost of technology is an important part of this phenomenon but doesn't explain all of it. Also crucial is that capital is much more abundant today, and because of low interest rates, it is much less costly than it was before. Similarly, not all data processing eliminates work; it also makes existing work more productive (or, as economists call it, "labor augmenting"), so labor benefits more from technology than capital does. Taken together, these factors may explain why labor share is decreasing more slowly than capital share.

But if both labor share and capital share are decreasing (albeit at a different pace), which is the big winner? Where is all the remaining income accumulating? Barkai's answer is that it has accrued as drastic increases in profits—or, in the language of economists, unwarranted markups for products and services. This would point to huge inefficiencies in markets and a lack of competition. And it would imply that investors (including everyone who has saved for retirement) were shortchanged as well. Share prices of companies may have risen over recent decades, but when factoring in low interest rates, the overall return that investors received was far below what it should have been. It would imply that *neither* workers *nor* investors have been fairly compensated, and that consumers have paid too much for the goods and services they purchased. And as critics have argued for years, the pockets of overcompensated executives have been lined at their expense. Even worse, related research indicates that as profits have soared, innovative activity and business dynamism have declined, at least in the United States.

A group of researchers headed by MIT economist David Autor has shed even further light on the underlying dynamic. They show that firms aren't winning across the board by increasing their profits. Rather, a particular type of firm is reaping unprecedented profits far exceeding what would be expected in an environment of efficient competition. Autor and his colleagues call these winners "superstar firms." They are often situated in "winner-takes-most" markets with strong network and feedback effects that lead to significant market concentration, and they have mastered the use of technology so completely that they can achieve very high revenues with comparatively limited labor and capital cost. Google, Apple, and Facebook are obvious superstars, but such firms exist in many sectors all over the world. Think of Spotify in Europe, online marketplace Alibaba and network technologies manufacturer Huawei in China, and tech giant Samsung in South Korea.

It's important to understand that the profits these superstar firms earn aren't fully reflected in their formal profit-and-loss statements. On paper, they may not report very high, formal profits. In part, this is because of creative corporate tax planning, but often it also reflects the fact that these superstars invest huge amounts every year in research and the development of new products and lines of business. Take Amazon as a case in point: in 2014 alone, it reported technology costs primarily for research and development of $9.3 billion. That Amazon can invest such huge amounts reflects the heavy relative markup it achieves on its main line of business. To some extent (but by far not fully), the stock market has realized the disparity between formal reported profits and the actual capacity of the superstars to achieve huge markup profits, and it shows in the trajectory of share prices: in 2015, six stocks accounted for almost all the gains of the NASDAQ Composite Index and five (Amazon, Google, Apple, Facebook, and Netflix) of them are obvious superstars.

This, however, is not evidence of a resurgent rise of the firm but rather a subtle sign of its demise. The traditional firm is a social construct designed to coordinate human activity—successfully getting lots of people to work together in a single organizational unit. By contrast, these superstar firms have a relatively small workforce; they thrive on automation, including for decision assistance. As entities with significant market power, many of them can exploit global regulatory competition wherever possible. To put it bluntly, organizationally these "superstars" succeed because they have streamlined internal coordination as much as possible while using trade and tax regulations to shelter themselves from the harsh winds of competition. It isn't too much of a caricature to say that they are turning into legal containers for the accrual of profits rather than large organizational units coordinating human activity.

As we replace the image of a data-fueled labor-destroying resurgence of finance capitalism with a more nuanced, but at least as troubling, view of skyrocketing profits, especially among tech-savvy superstar firms, we must substantially adjust our policy measures, both the distributive and the participatory perspective.

ON THE DISTRIBUTIVE SIDE, NEW AND INCREASED TAXES on capital (including a tax on machines) would essentially accelerate and deepen the decrease of capital share while leaving untouched the huge pile of corporate profits that lies at the root of the problem. That would be close to a textbook example of a bad idea.

A far more appropriate distributive policy response to skyrocketing profits accruing to certain firms is to tax these firms. This, of course, is not a new idea at all; it is the law of the land. In fact, at just shy of 40 percent, the United States nominally has one of the highest corporate tax rates in the world. The trouble is that none of the superstar firms actually has its profits taxed at that rate. In practice, the tax levied on America's largest corporations is less than half the nominal rate, with a number of them paying no tax in the United States at all.

This scandalous situation angers the public and has led many politicians on both sides of the aisle to advocate for tax reform. Unsurprisingly, there is little consensus about the details of such reform. Some suggest that tax authorities need to be tougher on corporations, forcing them to close the complex legal loopholes that reduce their tax liabilities and repatriate profits earned in tax havens abroad. Others focus on the relatively high nominal tax rate and suggest that lowering the nominal rate would be the strongest incentive to get corporations to comply.

Regardless of one's political predilections, it is very clear that from a distributive perspective, the soaring profits enjoyed by some US companies need to be balanced by a substantial rise in tax receipts from these corporations. That is what's needed as both labor and capital shares decline. The challenge is to push an increase in corporate taxation through a political system in the United States that too often is beholden to special interests.

The situation isn't radically different in Europe, either. Even though corporations, relatively speaking, pay more in corporate taxes than in the United States, experts have estimated that creative shifting of corporate profits to favorable tax regimes amounts to over $200 billion in lost tax revenues annually.

Economists have pointed out that large profits may be less worrisome if they are channeled back into the economy as investments. There has been a move to implement a so-called progressive consumption tax (PCT) to replace the progressive income tax. With it, personal income would be taxed only to the extent that income isn't reinvested. In the United States, a PCT enjoys support across the political spectrum, but that is no guarantee it will be legislated into existence anytime soon. And remember that in general, investments don't pay high returns anymore, so don't expect a raft of new investments.

If we accept that a general shift in the economy from reliance on money to reliance on rich data is underway, we ought to think more creatively about how to get companies to pay the taxes they owe. Governments might consider a partial payment of taxes in data rather than money. Car manufacturers might provide the public with anonymized sensor data from their cars; government could use it to identify particularly dangerous spots on the roads, thereby improving safety. A similar approach could improve food safety by using feedback data collected from farms and supermarkets. Feedback data from online learning platforms could help improve decision-making in the public-education sector, and decision-assistance data used for transaction matching could be reused in an early warning system that better predicts market bubbles.

Together with the data-sharing mandate we propose, this would make data available to small firms, especially start-ups, so that they can compete against the big players. It may also be a good way to jump-start innovation. The data could also be used by government to improve its services. And it might be offered to nonprofits, researchers, and society at large so that everyone can benefit from the profits of superstar firms.

The idea of some mandated disclosure of valuable information isn't new at all. In fact, the patent system is based on it: the privilege of a patent is only granted to those who reveal the workings of their inventions in their applications (at least that is the principle), so that everyone can learn from it, knowledge spreads, and after the expiration of a patent, others will have access to the invention for their own uses. Making valuable data available through tax payments would fulfill a similar purpose as firms shift from generating new ideas to generating data-driven feedback loops that improve products and services.

Unlike a conventional tax, a tax in data may be less onerous for those who must "pay" it. In contrast to money, data is non-rivalrous. It isn't used up, and a firm can make use of its own data even if it allows others to use it as well. Data taxes might be particularly of interest to start-ups that could use the money they don't have to pay in tax to grow their businesses. A data tax isn't particularly distributive, because both sides benefit, but it may be more welfare-enhancing overall as data is reused by multiple parties, and thus utilized more efficiently.

As firms would only be permitted to pay a fraction of their tax liability in data (not least because government needs money to fulfill its role), even a modest amount of a firm's annual tax bill paid in data could make a sizable difference. A resulting boost in innovation would also lead to new economic growth, taxes on which would mostly be paid the traditional way: with money.

If taxes paid in data make huge amounts of data available to the economy and society at large, this may signal what open-data proponents have long dreamed of but haven't yet achieved. The conventional conception of open data—making data held by government available to the general public—was limited by the minimal commercial and societal value of government data. The data that businesses are already transforming into value, on the other hand, may be more immediately useful.

On the participatory side, policy makers should consider not just supply-side strategies, such as (re)training workers, but demand-side measures as well. This isn't a recommendation for massive government-sponsored public works projects, but there are some smart incentives for stimulating demand.

One is to give firms a tax credit for each job they create. This would make it cheaper for companies to employ people. The tax credit would be offered every time a person takes a new job. It would have no direct effect on wages, but by growing demand for labor in general it would, albeit indirectly, facilitate wage increases.

A human employment tax credit is not a Luddite policy. It doesn't target firms on the verge of automating white-collar work, and the goal is not to dissuade these or other companies from automating. Rather, the policy would support the development of business models focused on innovative human services. As such, the policy is designed to act as a catalyst for the transition of the labor market, to foster experimentation with new service offerings and new business models.

A long-term side effect of such a tax credit would be a modest increase in overall inefficiency, if automation gets a bit more expensive relative to human labor. But such inefficiency might be desirable if it slows or arrests the decline of labor share. Countries particularly worried about the employment effect of data and data-rich markets may choose this path. Fostering the creation of jobs that are less likely to be automated is necessary if we want to replicate the job-creation miracle experienced during the late stages of the Industrial Revolution and counter the slowdown of business dynamism and entrepreneurial innovation.

Such a tax credit might seem on the surface to burden high-tech industries. A closer look, however, reveals a more nuanced picture. Making automation more expensive than human labor does not eliminate the incentive to automate. It's just that the return on automation—the cost savings realized—must be higher than before.

Ironically, therefore, a human-labor tax credit may stimulate efforts to develop technical advances that offer substantially higher cost efficiencies. Sectors ripe for automation may actually become more automated as a result of the tax credit. As the saying goes, this is not a bug but a feature. While encouraging job creation that is insulated from automation (at least in the medium term), it would stimulate further automation in those areas where humans are already in danger of being replaced by machines.

And to the extent policy makers want to retain retraining and reskilling programs, these programs need to be designed so that they are eminently and swiftly adaptable. It is no longer sufficient to react to changes in demand for specific skills long after the fact; rather, reskilling programs need to fashion cutting-edge analytics of rich data, including data from large online talent markets such as LinkedIn, to spot changes in skill demand as they occur and reflect them in reskilling programs without undue time lags. In fact, we suggest that policies are needed to help companies embed reskilling programs deep into their internal organizational DNA, rather than relying on external, government sponsored retraining efforts.

These three policy measures—making the companies that capture the profits of the data age pay the ones getting uprooted as a result of it, ensuring that markets stay competitive and that society as whole benefits from data, and making human labor just a bit cheaper than machines—will safeguard that all can reap their share of the data dividend. These actions will help our society cope with the changes of data-driven adaptive automation.

THIS APPROACH IS BASED ON THE BELIEF THAT COMPETitive markets are the key foundation of a healthy and prosperous society. This stands in stark contrast to assertions by people such as high-tech entrepreneur and investor Peter Thiel, who wrote in the *Wall Street Journal* that "competition is for losers" and that "if you want to create and capture lasting value, look to build a monopoly." Thiel's answer makes sense—from the vantage point of a passionate monopolist. But Thiel is completely wrong when it comes to appreciating the basic economic principles of data-rich markets and their consequences. His recipe for success would lead to astronomic profits for a small number of corporations, a slowdown in innovation, and gigantic inefficiencies that will cost consumers, workers, and investors dearly. This recipe may align with the political rhetoric of the Trump administration, its protectionist impulses, and its desire to bolster the firm and weaken the market. But it fundamentally misjudges what the data age is all about: a move away from money and capital, an appreciation of capturing the richness of reality through the richness of data, an embrace of the market over the firm, and an exceptional opportunity to improve the human ability to coordinate.

As data wins over money and the market wins over the firm, policies that address the consequences of data-driven markets must acknowledge this shift. Hence our suggestion of a progressive data-sharing mandate and a data tax—the ability to pay some taxes with data. But the need to focus on matters other than money is particularly obvious in the realm of human labor. We are all aware that many jobs provide more than money. They offer the chance for social interaction and afford meaning to people—important elements of our human identity.

In the past, the bundle of benefits we call employment wasn't very flexible, and the monetary dimension prevailed. Salary issues were often the primary topic of labor disputes. Although we all need money to live, the question is whether the specific bundle of benefits we call a job has to remain fixed, or whether we could and should rethink it and rebalance its elements. As we transition to data-rich markets, it's only logical to go beyond just considering money when choosing work.

We should ask whether a job is meaningful, at an organization that treasures similar values, and whether it offers the opportunity for worthwhile social interactions with coworkers and business partners. An increasing number of firms that are competing for talent, in particular in the high-tech sector, have been focusing on the nonmonetary elements of employment when recruiting and retaining talent. But these activities often have not been articulated in a coherent fashion, giving prospective employees no obvious way to easily identify the most appropriate workplaces for them. More important, these measures are rarely reflective of a comprehensive and coordinated strategy to fully track, provide, adapt, and improve the bundle over time. As intangibles, such as meaning, experience, and identity, gain in importance when people choose jobs, the elements of a job's benefit bundle should become a crucial feature of an organization's overall HR strategy.

From the outset—the way an organization recruits, and how it utilizes online talent platforms—we need a better ontology to capture the elements of a job's bundle of benefits, moving beyond salary and money and embracing data-richness. But it must not end there. It's not sufficient to give employees the ability to work less than full time, or partially from home, without also providing them with clear career paths when they choose to break out of the conventional mold. Much more so than in the past, smart organizations will realize that the goal is to facilitate and manage flexibility in their human resources considerably beyond what's been done so far.

Key to the future of human work is unbundling "employment," much as we have unbundled the CD (and the LP before it) into individual songs and let listeners create their own evolving musical mixes. We need to define the elements of work and make them flexible enough to be recombined. Enabling organizations to lend such flexibility to scale will be no small feat, nor will it be easy to bring discoverability to the various work elements so that individuals really will be able to pick and choose. And for organized labor, such as unions, the challenge is going to be how to retain their role of collective intermediary as work experiences and employment bundles fragment.

Measures within firms to unbundle work benefits, however, only go so far. If we want to truly enable humans to exercise a much broader freedom of choice in configuring work, we may need to help them, at least to some extent, to move beyond money. Workers remain focused on their salary because they need the money to get by. This is where the core idea of a universal basic income can arguably have its biggest impact. A true basic income, covering most of an individual's daily expenses, would, as we have noted above, be far too costly in most contexts. But a partial UBI, providing everyone with, say $500 a month, could provide an extra bit of flexibility, especially for lower wage workers, that would enable them to choose the job they like rather than the one that pays the most, to choose less than full-time employment to spend time with family, volunteer, or follow their entrepreneurial dream. Studies have shown that such flexibility does not result in slacking, but often allows for activities that create disproportionate meaning and identity for an individual.

If a partial UBI allows people to work less, we'll be able to keep more people in the workforce, even after automation is in full swing. And a reduced UBI could potentially be financed without having to raise taxes beyond what's reasonable. The aim is not distributive justice, but individual empowerment—for many more people to choose work beyond pay. The idea is to devalue the relative importance of money in employment, and to enable a richer perspective on work that straddles many more aspects of human fulfillment than just what is represented by the monthly paycheck.

In the future, we may come to appreciate the rich diversity of preferences for what we want from work. And embedding this appreciation in the way we organize and structure employment may allow us to leave salary and money behind as the main determinants of work. This vision calls into question the usefulness of conventional policy proposals addressing the fundamental reconfiguration of work. Solutions focused largely on money cannot address the challenges of an economy *after* money, at least not by themselves. They will have to be complemented by measures that address the intangible benefits that work confers on the human soul. Here, too, we need to reflect data's richness in reconfiguring markets and give meaning to the richness of work.

– 10 –

HUMAN CHOICE

A S A YOUNG MANAGEMENT CONSULTANT AT THE PAR-thenon Group, Katrina Lake realized that conventional brick-and-mortar retailers failed to understand the wants and needs of their customers because they didn't make adequate use of data. Gifted with an ability to think analytically as well as an innate ambition, she decided to go to business school so that she could become "a CEO of a retail company and lead the industry in technology and innovation." She got into Harvard Business School in 2009 and set out to achieve her goal.

About a year after she had entered the MBA program, Lake started her own venture, initially named Rack Habit, using data and algorithms to help women dress well and save time to boot. Together with a partner (with whom she has since parted ways), Lake sent out invitations to her friends and their friends in the Boston area, asking them to fill out an on-line questionnaire about their shopping preferences. The two founders transferred data on fit and style from the question-naires onto spreadsheets, then created simple profiles of their new customers. Then they went shopping for them. Dresses, blouses, and skirts were delivered by mail. Customers could keep them or send them back—a benefit that today is standard for curated shopping services. According to *Fortune* magazine, Rack Habit was "constantly maxing out Lake's $6,000-limit credit card" and "made no money." Still, the continually grow-ing number of customers proved that Lake and her partner were onto something.

The trouble with shopping for apparel isn't that the right product—one that fits and matches a person's preferences—doesn't exist. The problem is that most people, especially indi-viduals with demanding jobs and young kids, don't know how to find it quickly and easily. Or, as economists would put it: the apparel industry has increased variety but not discoverability. What shoppers need is a smart intermediary that can find the right items for them. These intermediaries exist: they are called personal shoppers, but unfortunately for most of us, they are too expensive. Rack Habit's core business idea was to give customers a competent personal shopper they can afford. By proposing to use rich data to match her customers and their preferences with the hugely diverse supply of apparel on the market, Katrina Lake raised $750,000 in seed funding in 2011.

With her newly minted Harvard MBA in hand, Lake moved to San Francisco, changed the name of her company to Stitch Fix, and began playing moneyball for women's fashion. It wasn't meant to be minor league. In 2015, the company was valued at $300 million in a third round of financing, prompting the *New*

York Times to list it as a potential start-up "unicorn"—a company valued at $1 billion or more. By 2016, as Devin Wenig started his crash program to make eBay's marketplace data-rich, Katrina Lake joined the *Forbes* list of richest self-made women.

At first glance, today's Stitch Fix looks like a traditional online retailer with a twist. It sends boxes containing five items of clothing and accessories to its customers (now both women and men), who pay for the items they choose to keep. But a closer look reveals that true to Katrina Lake's original intention, Stitch Fix acts much more like an intermediary in the market for clothing than like a retailer. For each box sent to a customer, Stitch Fix is charging a "styling fee" of $20. The fee is waived when a customer decides to keep at least two of the five items in a shipment (and there is an extra discount if a customer keeps all five items). Stitch Fix does not price its items to sell; there are no end-of-season sales, no Super Cyber Monday or Black Friday offers to please the comparison shopper. Stitch Fix is a "post-price" retailer, targeting customers for whom it is more economical to pay for good matches than to waste precious time locating goods they want to buy. This wouldn't work in every market—getting a good deal on a car can save a person thousands of dollars and may well be worth the effort. But for clothes, as Stitch Fix shows, it's a model that can be successful.

To achieve service at scale, Stitch Fix analyzes rich and comprehensive data streams. By 2016, the company employed more than seventy data scientists on a team headed by chief algorithm officer Eric Colson. Colson ran data science at Netflix, one of the rich-data pioneers. As it turns out, though, picking the right clothes is much harder than suggesting films to watch. Stitch Fix employs vastly more sophisticated data analytics than the standard social-filtering recommendation engines ("people who liked this movie also enjoyed this one"). In addition to asking its customers about their preferences when they first sign up, the company uses algorithms to examine pictures that customers like on Pinterest and extract salient features that reveal preferences its customers might not themselves even realize they have. The system works in exactly the same way as the machine learning systems we described in Chapter 4: customers don't have to make their needs and wants explicit, because the systems learn from how humans interact with the world around them.

Feedback also plays a crucial role at Stitch Fix. To begin with, every item a customer returns generates data. But customers are strongly encouraged to comment on each item they receive and they can do so in plain English, which, with the help of natural-language processing software, further refines a customer's preferences. Stitch Fix is also developing its own line of apparel, which uses preference data in the design process.

Stitch Fix's simple secret is that it understands data-rich markets and the crucial role data plays in customer satisfaction. As they put it: "Rich data on both sides of this 'market' enables Stitch Fix to be a matchmaker, connecting clients with styles they love (and never would've found on their own)."

But there is one more key ingredient in the company's success: the human touch. Stitch Fix employs hundreds of personal stylists. Working part-time, at home, and across the country, these stylists have the final say in choosing the items that go to each customer. They are aided by Stitch Fix's machine learning systems and matching algorithms and have access to rich and comprehensive data about a customer's preferences, including how those preferences have changed over time. But the final choice is theirs. And they add a personal note to every shipment, suggesting ways in which new items might be combined with other items in a client's closet or enhanced with accessories.

This personal voice shows the customer that somebody cares. It also establishes and maintains a relationship between the customer and the stylist. That helps retain the customer—after all, it's harder to dump a human than a machine. And it increases the likelihood that customers will provide feedback. The personal note creates a social debt that most customers choose to repay.

Perhaps most important, however, is that human stylists are far more heterogeneous in their views and tastes than machines. A rich diversity of personal stylists creates a universe of vastly different approaches to style, resulting in better matches between customers and stylists that in turn translates into higher sales and improved customer satisfaction. For Stitch Fix, harnessing the richness of human styles and preferences is an essential element of its success, one that makes humans indispensable.

Stitch Fix disentangled the tedium of shopping from the pleasure of choosing, and we might expect to see its model copied in the brick-and-mortar world. Perhaps in the future, we'll delegate the task of replenishing our groceries to our machine learning systems while we continue to enjoy perusing a boutique or browsing the shelves of a bookstore for ourselves. In fact, as we begin to patronize stores solely for the experience of shopping, as we rediscover the tantalizing joys of touching and seeing and deciding, we might even be willing to pay a fee to the best of these stores for that privilege.

A good strategic move for a large offline retailer might be to acquire Stitch Fix or one of its competitors, as Nordstrom did when it acquired Trunk Club. Its retail store could evolve into an experiential environment where customers can browse, perhaps with their personal stylists (physical or virtual) in tow. The brick-and-mortar store then complements the online business and becomes a place of enjoyable engagement with the merchandise and perhaps even with other customers. The goal is not to buy but to browse, to indulge the senses, soothed by music and energized by free espresso. The one thing we won't see in these revamped stores is signage that calls our attention to price and discount information. Price will not matter most in these settings; we will count on decision-assistance systems, such as digital shopping agents, to think about it for us.

We humans are tactile creatures and we love to use our senses and engage with the world. But we also enjoy each other's company. Stitch Fix is superbly successful at injecting this human element into its customers' experience. It's only a matter of time before others embrace data-driven markets the same way.

DATA-RICH MARKETS ARE UPENDING ONE TRADITIONAL money-based market after another. They provide better matches, resulting in more-satisfied participants. But that is not all. Better matches translate into less waste: fewer goods end up with buyers who can't make full use of them because they were looking for something else. Superior coordination means less idling and fewer inefficiencies. And, as money is no longer the prime dimension through which we match in data-rich markets, it may enable us to act on and express our values with every transaction we make, letting us transact conscientiously far more often than we can today. Data-rich markets further a more sustainable, less wasteful economy, especially compared to conventional money-based markets, and their excesses of greed and gluttony.

We only have one earth, so we must carefully manage the resources we have at our disposal. The one abundance we have is informational, and as collecting, conveying, and processing data become easier and less expensive, we will have more of it to use. The future of our economy lies in the clever exploitation of our informational surplus, and data-rich markets are the mechanisms and the places where we can achieve this. When artificial intelligence and Big Data meet the social reality of human coordination, we can become more sustainable.

Spurred by "smart meter" technology, for example, energy markets will become data-rich, transitioning from their inefficient and fragile current state, in which a limited number of large producers provide energy for many, toward a much thicker market in which a huge number of diverse participants, including home-based producers of energy (think solar) and storage (think batteries), can better coordinate with each other. Not only will we waste less energy, this will enable us to more efficiently use the smart grid, an advanced energy distribution infrastructure.

Shipping logistics will benefit from data-rich markets as well. About one in four trucks drive empty because there is no efficient way for them to get freight for a particular leg of a trip. Self-driving trucks alone will not change this situation, but data-rich markets can provide better matches of trucks and freight. The effect of such enhanced coordination would translate into lower emissions and improved sustainability.

Or consider health care. In advanced economies, especially the United States, health care costs have reached unsustainable levels. Although we may live slightly longer, we spend an enormous amount of money in return for that small gain. In part, this is because we lump patients together and treat them alike; much like in money-based markets, we have reduced the amount of information we get so that it won't overwhelm us. But no tumor is exactly like another, and thus diagnosis and treatment, too, need to be individualized. And what's true for cancer applies to many other illnesses. Taking lessons from data-rich markets, in health care we, too, can leave behind oversimplification and make medicine more precise.

Schools could use rich data to improve the way they match pupils with teachers, learning materials, and pedagogical methods. Not all pupils are the same, and what works for one may not work for another. The principles that underlie data-rich markets can be applied more broadly and can lead to better outcomes while reducing wastage. We'll live better, more meaningful, and more sustainable lives.

BECAUSE DATA-RICH MARKETS ARE SO MUCH BETTER than traditional markets, they put tremendous pressure on us to readjust our concept of the firm. In the past, firms were organizational units of humans working together. Some firms may have used machines, but all firms were managed by humans. In the future, that may no longer be the case. We may encounter firms that employ many well-paid humans doing work only they can do, but those firms may be managed in substantial part by machines. And we will see other firms morph from social organizations into largely legal entities that reap profits but have dispensed with many human employees. The latter, the empty corporate shell, will be an artifact of corporate and tax laws enacted when the landscape of business looked much different. The former, the human-centric firm, is where adaptive systems mesh with those human qualities that cannot be easily automated. Like Stitch Fix, the human-centric firm isn't Luddite but rather enormously data savvy. It enhances the data-driven market rather than competing with it.

There are a great number of sweet spots along the spectrum between empty shell and human-centric organization that firms can occupy. Firms will need to identify their places along this spectrum and then change themselves to flourish there, whether they are Daimler or Spotify, eBay or Apple, Alibaba or Barclays, established global franchises like McDonald's, or recent fintech start-ups like Stash, or the little organic bakery across the street. Perhaps in a few decades we'll call this the Great Adjustment and be able to tell the story of how firms large and small changed in the wake of the Great Recession. It isn't a change most companies have been looking forward to; it certainly isn't one that many of them will initiate. But those firms that embrace this adjustment and transform themselves to become fundamentally more human will enjoy an unquestionable advantage.

In banking and the financial services industries, there is mounting pressure to further commodify financial services, especially around payment. New intermediaries have appeared on the scene that promise to facilitate and utilize richer and more comprehensive data streams, going far beyond our conventional obsession with money and price. Traditional players, such as large corporate banks, face a difficult choice. The path to commodification is one they know and rightfully dread. The path to turning into a successful data intermediary offers more upside but is also much harder, because data-richness isn't something many banks are good at. They would have to change their inner workings, and that's a challenge. "It's difficult to see the picture when you're inside the frame," in the words of the late Eugene Kleiner, one of the pioneer venture capitalists of Silicon Valley.

Take a good look at the marble palaces of money—the banks that dot the financial centers around the world. They telegraph a message of power and wealth to all who behold them, but by the late 2020s, many of them will be gone. Not because we abolished money but because we no longer need a crude symbol for a service that dispenses with detail and reduces informational richness. The new financial intermediaries will be either superefficient or rich in data—and perhaps even both—while the vestiges of old-fashioned finance capitalism will wither away, replaced by the new breed of data-capitalism and powered by data-rich markets.

The difference between a Silicon Valley angel investor and a European bank clerk extending a loan to an entrepreneur isn't money but the rich and valuable information for entrepreneurs that the angel provides. Since the early days of Georges Doriot's American Research and Development Corporation, the first modern venture capital firm, we have been calling VCs "capitalists," but this may long have been a misnomer. The best angels and VCs offer their clients vastly more than just a check. Perhaps it's time to call them *venture informers* and dispense with the term "capitalist."

We can take it further still: with the demise of the role of finance capital as the catalyst that enables firms to grow and dominate, it may be time to close the door on history and officially eliminate the term of "capitalism." It had a good run and has shaped our market economy over the last centuries. But we have an opportunity to fashion an alternative to finance capital as the central source of power in our economy. Instead of capital and firms, we can imagine data-rich markets that empower humans so that they can better work with each other.

DATA-RICH MARKETS AREN'T SOLVING ALL OUR PROBlems, nor are they without inherent structural weaknesses. The success of any data-driven market, therefore, hinges on its design—and on the rules under which it operates. The most critical element in designing data-rich markets is to protect them against concentration, not only at the level of participants (the classic one-seller-many-buyers and one-buyer-many-sellers situations) but also at the level of decision processes. If many or all of the adaptive machine learning systems we use to help us make transaction decisions have the same flaw, the market as a whole becomes vulnerable. That's why a deep diversity of decision-assistance systems is indispensable. The progressive data-sharing mandate—the mechanism we have suggested to ensure such diversity—is designed to protect not just against the concentration of adaptive systems but also, through the sharing of different data subsets, to guard against all competitors using the same data input to build their systems. The key is to foster a diversity that continues to translate into robust competition.

When it comes to the broader social implications of data-driven markets, we abhor the gleeful optimism of the techno-utopians as much as we shun the gloom of the perennial doomsday prophets. Rather than pretending to predict the future, we should prepare ourselves to shape it, readying the right levers and mechanisms so that we can stimulate beneficial dynamics and mitigate negative consequences whenever and wherever they arise. Sometimes this necessitates adding novel items to our existing policy tool kit, such as (data) taxes paid in data or a recurring human-labor tax credit. And sometimes it requires a different perspective, such as disentangling the notion of work from the notion of pay.

When money and price prevailed, it may have been appropriate to understand labor in terms of wages and financial benefits. But as we move to data-richness in markets, we need to move beyond salary figures when we debate the merits of human labor. When we accept that work is more than a job that pays the bills—that for many people it offers identity, community, and a sense of belonging—we will also realize that helping people find good work and keep it will continue to be a central role for society, even as machines take on an increasing number of responsibilities. The task of the market is to be efficient, and with data-richness, better matches will lead to vastly improved efficiency (and sustainability). By contrast, our task as humans is not to be most efficient but to be truly human, from being creative and adventurous in our thinking about the new to engaging with each other and forging meaningful social bonds.

In a world of more and more machines, what will remain for us humans to do? Are we the dinosaurs of the data age? Will we find ourselves confined to a reservation, much to the amusement of the machines that call the shots? Even in an era of data-richness, we remain convinced that humans will continue to be in the lead, if they want to be. Rich data will permit us to choose which decisions we want to make ourselves, and to leave the rest to adaptive systems that are aware of our preferences to find the best match in the market.

By freeing our minds from routine decisions, we can focus on decisions that really count, and those that we enjoy making. Eventually we may even be able to delegate to adaptive systems some of the decisions we fret over—the decisions we worry we won't get right because of our biases; the ones we are apprehensive about because we don't know enough and don't have the time to inform ourselves. And we'll be able to tell our decision-assistance systems how much "correcting" we'd like them to do. It won't be a simple binary either-or. Rather, we'll be able to dial up (or down) the amount of help we'd like. We'll choose to choose.

Experts have warned that Big Data and artificial intelligence may endanger human volition by making decisions not only about what we buy, but with whom we coordinate. The fear is that as we delegate decisions to machines, we give up an important freedom to shape our social sphere. But rather than taking away the freedom to choose, data-rich markets empower us, as we explained in Chapter 4, to spend our time and energy on being human.

An actual, immediate choice about which decisions to make will force us to face questions about how we choose. Which decisions should we reserve for ourselves and which should we delegate? If data-driven adaptive systems will offer us better answers to questions such as which school we should send our kids to or which hospital an ambulance should take us to in case of an emergency, then should we delegate that decision to the machines or retain it as the exclusive province of human responsibility? What are we aiming for in decisions, anyway—getting the correct answer or the one that makes us happy (after all, we, not the machines, must live with the consequences)? Until now we rarely faced such choices, but in the future we routinely will. Developing a good, solid sense of how to choose is a core competency we'll have to develop and maintain.

This ability to choose what to choose is fundamentally empowering to humans. It preserves our chance to contribute to the fate of the universe and may ensure us an enduring seat at the table of evolution. But it is also a novel challenge that comes with responsibilities. Choosing means picking one option over another, and forgoing having it all. Data-rich markets are amazing instruments of choice—they will help us choose well. But they will not absolve us from the need, ultimately, to select.

Losing the individual freedom to choose in the name of security, simplicity, coherence, or perhaps just plain old maximization of profit would be a terrible loss, far beyond the economic inefficiencies it would cause. It would erode and then abandon a core principle of every free society. That is why we need to be alert for proposals that tout the dangers of individual choice and advocate the prudence of centralized power. It is why we need to be watchful about data-driven feedback effects, just as we need to resist impulses of a strong controlling government.

There is yet another enemy of data-rich markets and individual choice. It is the vision that humanity will soon overcome resource scarcity, and the belief that machines, and their seemingly infinite ability to accomplish complex tasks at low or practically no cost, will recycle the resources we have forever, essentially taking us into a true utopia. In it, humans, freed from daily chores, will enjoy life and have the means to live it to the fullest. The end of scarcity has been predicted before, perhaps most vocally in the 1970s by conservative economist Julian Simon. Now an expanded version of the idea is resurfacing. Erik Brynjolfsson, a business school professor at MIT's Sloan School of Business and coauthor of an influential book about the consequences of artificial intelligence on human labor, seems sympathetic. "A world of increasing abundance, even luxury, is not only possible, but likely," he suggests.

Proponents of this view use the term "fully automated luxury communism" (conjuring images of Leonid Brezhnev wearing Gucci loafers) for the idea that we all can work less and still enjoy whatever we want. The writer Aaron Bastani, who claims to have coined the term, believes it would lead to "Cartier for everyone, MontBlanc for the masses and Chloe for all." The British newspaper the *Guardian* puts it a bit less materialistically: "Humanity would get its cybernetic meadow, tended to by machines of loving grace." In such a world, the argument goes, we would no longer have to choose; we could have the cake and eat it, too. And the market would become obsolete, like an eight-track player or ashtrays on restaurant tables.

Advocates of abundance have built that vision on a fundamental fallacy. They focus on physical resources and misunderstand that the market is not just a way to allocate scarce physical goods, but a way for humans to coordinate with each other effectively and efficiently. As humans coordinate, we face the scarcity of time. Without perfect immortality, we need markets to coordinate with each other in the limited time we have.

Rather than hoping for a timeless world that does not exist, we need to embrace the reality we live in. We do not dispute the existence of severe economic woes throughout the world. We, too, see them as acute challenges. But the market isn't the root cause of the problems. Instead, we view it as our best hope for overcoming them. We suggest that what matters is the ability to coordinate with each other in time to reach goals we could never achieve alone.

We are convinced that the market is here to stay. Our future as humans is not about centralizing power, nor is it about consumption and opulence; rather, it is about interacting with each other so that we can spend the very resource that we cannot replenish—our time—most meaningfully. But a comprehensive and sustainable freedom to choose requires not only the shift from money to data and the technical advances we have detailed in Chapter 4.

It rests on the market as a decentralized social mechanism coordinated by humans. Eliminate the decentralized market, and the empowering quality of data vanishes. That is why we call the shift from money to data a revival of the market instead of the rise of artificial intelligence or the advent of Big Data. Without the market, neither data nor technology will protect—let alone advance—humankind and help people work together. Hence, in this book, the market has taken center stage. Data (and technology) are merely enabling its renaissance.

The revival of the market is founded on access to rich and comprehensive data and the ability to translate that data into decisions. In turn, this lets us outgrow condensed and simplified information that has ruled traditional markets for centuries. Simplifying reality so that we are more comfortable with it is an old human strategy. It's useful when we don't know better, when we don't have ample information; and when we can't comprehend well, when our mental faculties fail us and we lack the tools to acquire a more accurate, detailed perspective.

ASSUMING THE EARTH WAS FLAT WAS A SIMPLIFICATION we employed for centuries because it worked—until we needed to progress further. We replaced it with something more complex—a globe rather than a flat plain—but the complexity helped us advance. We are doing the same as we transition from money-based to data-rich markets.

Fundamentally, this shift is part of a larger, broader movement that began hundreds of years ago. It drove Francis Bacon to emphasize the need for empirical evidence and René Descartes to look for reasons. It prompted Immanuel Kant to suggest that reason links to morality and Adam Smith to examine the power of market coordination. It led Hannah Arendt to look at the nature of power and John Rawls to ponder justice. It has propelled humans along the path of knowledge and given us insight into the world we live in—a world rich in information, more colorful, more diverse, more nuanced, and more exciting than we realized. This voyage isn't over; it will continue.

It's important all of us appreciate that we cannot dumb down reality to make it conform to our cognitive limitations; that when we limit the possible explanations of how the world works to the simple one, the one that's easiest to grasp, or the one that we have always believed in, we confine our imagination, and we constrict our understanding of the world to the obvious. This may have been a suitable approach when we lacked an alternative, but with data we no longer must.

Humanity's future is going to be one of knowledge and insight—if we want it to be. This will mean leaving behind many of the simplifications we trusted and embracing the world in all its diversity. Data alone won't do that: to open the window to new insights, we need to open our minds. But contrary to dystopian predictions, to the view that data is cold—a view that pits technology against humans—we believe that, through data-richness, our future is going to be profoundly social and thus deeply human.

ACKNOWLEDGMENTS

"**D**on't tell me what I already know," advised Lewis Branscomb two decades ago. This sound advice guided us in writing this book. Of course, many of our readers will know at least some aspects of the story we tell. But our hope is that the overall narrative we offer is new to most of our readers, and that it will inspire them to look differently at markets and money, firms and finance, and digitization and data.

The argument we put forward is neither easy nor simple; it requires effort to consider it—to put aside, at least for a bit, conventional beliefs and long-held convictions about how economy and society work, and to appreciate an alternative perspective.

Over the last twenty-four months, we had countless conversations with experts and colleagues around the world, on every aspect of our story. Learning from them has been the high point of our journey. They gave generously of their time and offered invaluable insights. Many of them spoke openly but requested that we not mention their names. Some of the ones we are allowed to thank in public include venture capitalist Alfred Wenger; blockchain enthusiasts Don and Alex Tapscott; data analyst Maximilian Eber; privacy regulator Yann Padova; insurance insider Christian Thiemann; economist Simcha Barkai; poker pros Jason Les and Don Kyu Kim; ontology authority Madi Solomon; pricing expert Florian Bauer; computer scientists Manfred Broy and Johannes Buchmann; entrepreneur August-Wilhelm Scheer; HR visionary Thomas Sattelberger; Mattias Arrelid and Anders Ivarsson at Spotify; decision expert Francis de Vericourt; mathematicians Erich Neuwirth and Max von Renesse; transportation design expert Stephan Rammler; philosopher Christoph Hubig; polymath Heinz Machat; as well as journalist colleagues Ludwig Siegele, Uwe Jean Heuser, Wolf Lotter, and Christoph Koch.

We would like to thank in particular our agent, Lisa Adams of Garamond Agency, who has always been there for us, believing in and championing the book from the beginning. We also thank our editor at Basic Books, TJ Kelleher, who has challenged us to focus on clarity, consistency, and concept. We acknowledge the help of Robin Dennis at the proposal and early drafting stage, Barbara Clark for copyediting an early version and Michele Wynn for copyediting the final version of the manuscript, as well as Phil Cain for his excellent checking of facts.

We have benefited from institutional support in writing this book: Viktor gratefully acknowledges a sabbatical from Oxford University. Thomas would like to thank Karl Neumar and Johann Blauth for exempting him from all duties at QuantCo during the months of intense writing and *brand eins* magazine for granting the strong support and generous freedom he enjoys as its technology correspondent.

Writing a book is always an undertaking that ends up consuming more time and effort than originally envisaged. But this book repeatedly pushed us to the limits—to think deeper and try harder. Therefore, far beyond the bow that is customary, we thank our families for their patience.

NOTES

CHAPTER 1: REINVENTING CAPITALISM

1 **the online marketplace's twentieth-anniversary event:** Marco della Cava, "EBay Turns Twenty with Sales Plan Aimed at Rivals Like Amazon," *USA Today*, September 16, 2015, http://www .usatoday.com/story/tech/2015/09/16/ebay-turns-20-sales-plan -aimed-rivals-like-amazon/72317234.

1 **traded on eBay's platform:** "EBay: Twenty Years of Trading," *Economist*, September 3, 2015, http://www.economist.com/blogs /graphicdetail/2015/09/daily-chart-1.

1 **active eBay users:** Leena Ro, "For eBay, a New Chapter Begins," *Fortune*, July 19, 2015, http://fortune.com/2015/07/19/ebay -independence.

2 **"a general rallying the troops":** Della Cava, "EBay Turns Twenty."

2 **"due for a reset":** Ibid.

2 **eBay's recent troubles:** Nicole Perlroth, "EBay Urges New Passwords After Breach," *New York Times*, May 21, 2014, https://www.nytimes.com/2014/05/22/technology/ebay-reports-attack-on-its-computer-network.html?_r=0.

2 **sellers of Yahoo's shares:** Matt Levine, "How Can Yahoo Be Worth Less Than Zero?" *Bloomberg*, April 17, 2014, http://www.bloomberg.com/view/articles/2014-04-17/how-can-yahoo-be-worth-less-than-zero; see generally, Richard H. Thaler, *Misbehaving: The Making of Behavioural Economics* (London: Allen Lane, 2015), 244–253.

7 **"electronic markets":** Thomas W. Malone, Joanne Yates, and Robert I. Benjamin, "Electronic Markets and Electronic Hierarchies," *Communications of the ACM*, June 1987, https://www.researchgate.net/publication/220425850.

8 **a staggering 9,300 percent upturn:** "The Zettabyte Era: Trends and Analysis," Cisco White Paper No. 1465272001812119, June 7, 2017, http://www.cisco.com/c/en/us/solutions/collateral/service-provider/visual-networking-index-vni/vni-hyperconnectivity-wp.html.

CHAPTER 2: COMMUNICATIVE COORDINATION

18 **each new foursome got into place:** Efren Garcia, "Historic Record in Catalonia's Human Tower Building," *Ara: Explaining Catalonia*, November 23, 2015, http://www.ara.cat/en/Historic-record-Catalonias-tower-building_0_1473452720.html; see also the video of the tower being built on YouTube, https://www.youtube.com/watch?v=qTP-Xp7v6m0.

18 **a girl had fallen to her death:** "Una niña doce años muere al caerse de un 'castell' de nueve pisos en Mataró," *Libertad Digital*, August 4, 2006, http://www.libertaddigital.com/sociedad/una-nina-de-12-anos-muere-al-caerse-de-un-castell-de-nueve-pisos-en-mataro-1276285054.

19 **set a new world record:** Stefania Rousselle, "Building Human Pyramids for Catalonia" (video "Climbing for Catalonian Pride"), *New York Times*, November 7, 2014, https://www.nytimes.com/video/world/europe/100000003222118/catalonians-climb-high-to-exhibit-pride.html.

19 **"no limits but the sky"**: Here it seems exactly right—but also some-what wrong—to quote Castilian Miguel de Cervantes's *Don Quix-ote*, pt. 3, chap. 3.

19 **complexity of the structure is the principal concern:** A *castell* with an *aguila*, or needle—a tower of single individuals standing one atop the other that is revealed as the main castle is deconstructed—is the most complex and coveted form.

19 **has come to mean "working together":** UNESCO, "Human Tow-ers," YouTube video, November 5, 2010, https://www.youtube .com/watch?v=-iSHfrmGdyo.

21 **became possible to protect a dependent child:** Sarah Blaffer Hrdy, *Mothers and Others: The Evolutionary Origins of Mutual Understand-ing* (Cambridge: Harvard University Press, 2009).

21 **opened the floodgates to globalization:** Lloyd G. Reynolds, "Inter-Country Diffusion of Economic Growth, 1870–1914," in Mark Gersovitz, Carlos F. Diaz-Alejandro, Gustav Ranis, and Mark R. Rosenzweig, eds., *The Theory and Experience of Economic Devel-opment: Essays in Honor of Sir W. Arthur Lewis* (New York: Rout-ledge, 2012), 319.

21 **in exchange for a freshly transcribed copy:** Mark Kurlansky, *Paper: Paging Through History* (New York: W. W. Norton, 2016), 13.

21 **revolutionary eighteenth-century *Encyclopédie*:** Ibid., 231. See also: Frank A. Kafker and Serena Kafker, *The Encyclopedists as Individuals: A Biographical Dictionary of the Authors of the Encyclopédie* (Oxford: Voltaire Foundation, 1988), http://encyclopedie.uchicago.edu.

22 **40 million articles in nearly three hundred languages:** These numbers are from *Wikipedia*'s home page as of December 2016.

22 **system to classify the planet's life forms:** Wilfrid Blunt, *Linnaeus: The Compleat Naturalist* (Princeton: Princeton University Press, 2002), 185–193.

22 **led to the theory of evolution:** Roland Moberg, "The Development of Protoecology in Sweden," *Linné on Line*, University of Uppsala, 2008, http://www.linnaeus.uu.se/online/eco/utveckling.html.

22 **The moon landing required:** "Apollo 11 Mission Report," NASA, n.d., http://www.hq.nasa.gov/alsj/a11/A11_PAOMissionReport .html.

22 **10,000 scientists from over one hundred countries:** Roger High-field, "LHC: Scientists Jockey for Position in Race to Find the

Higgs Particle," *Telegraph*, September 10, 2008, http://www
.telegraph.co.uk/news/science/large-hadron-collider/3351478
/LHC-Scientists-jockey-for-position-in-race-to-find-the-Higgs
-particle.html.

23 **"looks with grace upon each link":** Quoted in Moberg, "The Development of Protoecology in Sweden."

23 **"Coordination ranges from tyrannical to democratic":** Charles E.
Lindblom, *The Market System: What It Is, How It Works, and What
to Make of It* (New Haven: Yale University Press, 2002), 20.

25 **the word *economics*—the Greek *oikonomia*:** Dotan Leshem,
"Retrospectives: What Did the Ancient Greeks Mean by *Oikonomia?" Journal of Economic Perspectives* 30, no. 1 (Winter 2016), 225–
238, https://www.aeaweb.org/articles?id=10.1257/jep.30.1.225.

26 **In a market, coordination is decentralized:** Of course, not every market is fully decentralized in practice. For instance, if there's
only one buyer and many sellers, then the decision to buy is done
by a single participant. These markets are often characterized as
suffering from concentration. But sometimes markets are deliberately designed to have a central decision-making entity; the market
to assign doctors to residency programs in the United States is a
prime example. As we shall see, often these special markets cannot
utilize price to convey information and centralize decision-making.

26 **not just on the level of a household:** Lindblom, *The Market System*, 5.

27 **a factor of almost 2,000 since the 1500s:** There is significant debate about the accuracy of historical figures, often estimations based
on many assumptions. We also equated market volume with the
gross global product at purchasing-power parity—that, too, is quite
an approximation. See J. Bradford DeLong, "Estimating World
GDP, One Million B.C.–Present," http://holtz.org/Library/Social
%20Science/Economics/Estimating%20World%20GDP%20by
%20DeLong/Estimating%20World%20GDP.htm; for recent
world GDP figures, we used the data from the CIA Factbook
(https://www.cia.gov/library/publications/the-world-factbook
/geos/xx.html).

28 **100–200 million firms that exist:** Most nations do not track the
number of firms, so no global figure exists; the best we can do is to
go with estimates based on total employment and employment size;

see here for some estimation approaches: http://www.quora.com /How-many-companies-exist-in-the-world.

28 **employment by private-sector firms in high-growth countries:** Jin Zeng, *State-Led Privatization in China: The Politics of Economic Reform* (London: Routledge, 2013), 28–29 and 52–53.

28 **In the developed . . . (OECD) nations:** See http://www.oecd -ilibrary.org/sites/gov_glance-2015-en/03/01/index.html ?contentType=&itemId=%2fcontent%2fchapter%2fgov_glance -2015-22-en&mimeType=text%2fhtml&containerItemId=%2 fcontent%2fserial%2f22214399&accessItemIds=.

29 **implementing the "five-dollar day":** M. Todd Henderson, "Everything Old Is New Again: Lessons from Dodge v. Ford Motor Company," John M. Olin Program in Law and Economics Working Paper No. 373, University of Chicago Law School, 2007, 2–13, https://papers.ssrn.com/sol3/papers.cfm?abstract_id=1070284.

30 **shareholders demanded a larger dividend:** The case was *Dodge v. Ford Motor Company* (1919). See ibid.

30 **information should flow upward:** Henry Ford, *My Life and Work* (Garden City, NY: Doubleday, Page, 1922).

30 **firms will increase in size and combine:** The Marxist *Monopoly Capital*, written in the 1960s, is perhaps one of the most cited critiques from the left, although the authors' argument against "monopoly capitalism" echoes Lenin's earlier work; Paul A. Baran and Paul M. Sweezy, *Monopoly Capital: An Essay on the American Economic and Social Order* (New York: Monthly Review Press, 1966). The economist and admirer of innovation Joseph Schumpeter was more nuanced in his critique: on the one hand, he identified large firms as surprising places of innovation; on the other hand, he worried that capitalism may become undone as monopolies cripple the human urge to innovate; see generally Thomas K. McCraw, *Prophet of Innovation* (Cambridge: Harvard University Press, 2007).

30 **firm hasn't yet replaced the market:** For data on the growth of large firms, and especially the Fortune 500's increasing share of US GDP over time (growing from 58 percent in 1994 to 73 percent in 2013), see Andrew Flowers, "Big Business Is Getting Bigger," *FiveThirtyEight*, May 18, 2015, http:/fivethirtyeight.com/datalab /big-business-is-getting-bigger.

31   **buy and assemble parts made by others:** John Hagel III and John
     Seely Brown, *The Only Sustainable Edge: Why Business Strategy De-
     pends on Productive Friction and Dynamic Specialization* (Cambridge:
     Harvard Business School Press, 2005), 106–109.

31   **the assemblers broke down the design:** Dongsheng Ge and Taka-
     hiro Fujimoto, "Quasi-Open Product Architecture and Techno-
     logical Lock-In: An Exploratory Study on the Chinese Motorcycle
     Industry," *Annals of Business Administrative Science* 3, no. 2 (April
     2004), 15–24, http://doi.org/10.7880/abas.3.15.

32   **dramatically expanding the number of market participants:**
     K. Yamini Aparna and Vivek Gupta, "Modularization in the Chi-
     nese Motorcycles Industry," IBS Center for Management Research,
     Hyderabad, India, Working Paper BSTR/165, 2005, http://www
     .thecasecentre.org/main/products/view?id=66275, 5–7.

32   **Honda's sales fell from 90 percent:** Hagel and Brown, *The Only
     Sustainable Edge*, 108–109.

CHAPTER 3: MARKETS AND MONEY

37   **market volatility plummeted:** Robert Jensen, "The Digital Pro-
     vide: Information (Technology), Market Performance, and Wel-
     fare in the South Indian Fisheries Sector," *Quarterly Journal of
     Economics* 122, no. 3 (August 2007), 879–924, https://academic
     .oup.com/qje/article-abstract/122/3/879/1879540/The-Digital
     -Provide-Information-Technology-Market.

39   **"The market is essentially an ordering mechanism":** Friedrich
     August von Hayek, "Coping with Ignorance," Ludwig von Mises
     Memorial Lecture, Hillsdale College, Hillsdale, MI, July 1978.

40   **the market for used cars:** George A. Akerlof, "The Market for
     'Lemons': Quality Uncertainty and the Market Mechanism," *Quar-
     terly Journal of Economics* 84, no. 3 (August 1970), 488–500, http://
     qje.oxfordjournals.org/content/84/3/488.short.

40   **fewer peaches are offered for sale:** Information asymmetries can
     also cause sellers to lose out when they undervalue their goods and
     services and a more informed buyer takes advantage of it. For exam-
     ple, a seller may offer an initial service at a loss to a buyer, imagin-
     ing that the transaction will lead to repeat sales, not knowing that

the buyer never intends to come back—or would only do so for the same low price. The seller's "loss leader" leads to nothing but a loss.

41 **before copycats appear and free ride**: In his famous book *The Theory of Economic Development*, Joseph Schumpeter argued that entrepreneurs, by definition, have discovered a category of exclusive information. They're the first people to identify a new market, patent an invention, launch an efficient means of production, or introduce some other "new combination"—a way to coordinate human activity—before anyone else is aware of it. For Schumpeter and his acolytes, the resulting information asymmetry creates an economic incentive. Even though the market becomes less efficient, that is the price we pay for innovation. Hence, information imbalances aren't necessarily bad—up to a point. The incentives created by information asymmetries are essential to innovation, but the reward for innovation cannot be permanent without harming the market. Asymmetries must be fleeting and temporal lest market predators exploit their information monopoly forever, creating a vacuum, a black hole in which information is trapped. Information vacuums force buyers to make suboptimal decisions. Thankfully, most information asymmetries are temporary. Competitors copy or emulate innovations and catch up, erasing the innovator's information advantage.

41 **a transaction takes place that shouldn't**: There are numerous such cases; see, e.g., the discovery of an original Declaration of Independence hidden in a picture bought at a flea market (Eleanor Blau, "Declaration of Independence Sells for $2.4 Million," *New York Times*, June 14, 1991, http://www.nytimes.com/1991/06/14/arts /declaration-of-independence-sells-for-2.4-million.html).

42 **German pharmaceuticals company, Grünenthal**: Grünenthal was the first West German company to produce and sell penicillin after the occupying forces lifted their ban on the drug's production in the country. See https://en.wikipedia.org/wiki/Grünenthal_GmbH.

42 **By mid-November, he informed Grünenthal**: http://www .contergan.grunenthal.info/thalidomid/Home_/Fakten_und _Historie/342300049.jsp?naviLocale=en_EN.

42 **the last British "thalidomide baby"**: Nick McGrath, "My Thalidomide Family: Every Time I Went Home I Was a Stranger," *Guardian*,

August 1, 2014, https://www.theguardian.com/lifeandstyle/2014
/aug/01/thalidomide-louise-medus-a-stranger-when-i-went-home.

44 **juggle about half a dozen distinct pieces of information:** The
insight was originally made by psychologist George Miller (see
George A. Miller, "The Magical Number Seven Plus or Minus
Two: Some Limits on Our Capacity for Processing Information,"
Psychological Review 63, no. 2 (March 1956), 81–97, http://psycnet
.apa.org/psycinfo/1957-02914-001, and his article became one of
the most cited in the academic literature. More recent studies have
shown that the number of items isn't fixed, but that a human's working memory is a very limited resource that can be allocated in a variety of ways (see, e.g., Wei Ji Ma, Masud Husain, and Paul M. Bays,
"Changing Concepts of Working Memory," *Nature Neuroscience* 17
[2014], 347–356).

45 **"Money is the root of most progress":** Niall Ferguson, *The Ascent
of Money: A Financial History of the World* (New York: Penguin
Books, 2008), 4.

47 **"prices can act to coordinate":** "A Conversation with Professor Friedrich A. Hayek" (1979), in Diego Pizano, ed., *Conversations with
Great Economists* (New York: Jorge Pinto Books, 2009), 5.

47 **scholars of money point to numerous roles:** See, e.g., Nigel Dodd,
The Social Life of Money (Princeton: Princeton University Press,
2014), 15–48.

49 **so infatuated with markets and money:** For a discussion of the
limits of markets, see, e.g., Margaret Jane Radin, "From Babyselling
to Boilerplate: Reflections on the Limits of the Infrastructures of
the Market," *Osgoode Hall Law Journal* 54, no. 2, forthcoming; Osgoode Legal Studies Research Paper No. 28/2017 (January 24,
2017); University of Michigan Law and Economics Research Paper
No. 16-031; University of Michigan Public Law Research Paper
No. 530, https://ssrn.com/abstract=2905141.

50 **a better-than-even chance to know the truth:** Cass R. Sunstein,
Infotopia (New York: Oxford University Press, 2006), 25ff.

51 **the probability of events related to Google projects:** Other companies have also experimented with prediction markets, but Google's appear to be the largest and longest experiments conducted
in the corporate world. See Bo Cowgill, Justin Wolfers, and Eric
Zitzewitz, "Using Prediction Markets to Track Information Flows:

Evidence from Google," in Sanmay Das, Michael Ostrovsky, David Pennock, and Boleslaw K. Szymanski, eds., *Auctions, Market Mechanisms and Their Applications* (Berlin: Springer, 2009), 3, http://link.springer.com/chapter/10.1007/978-3-642-03821-1_2.

51 **first issue of the product-comparison magazine:** "Consumer Group Formed: New Organization Plans to Give Data on Goods and Services," *New York Times,* February 6, 1936, http://query.nytimes.com/gst/abstract.html?res=9F0CE0DF153FEE3 BBC4E53DFB466838D629EDE.

54 **Prices ending in nines:** Even if policy makers prohibit prices ending in nine, recent research has shown that the market adjusts quickly to the restriction and shifts from prices ending in ninety-nine to prices ending in ninety, with the same deceiving effect on consumers; see Avichai Snir, Daniel Levy, and Haipeng Chen, "End of 9-Endings, Price Recall, and Price Perceptions," *Economics Letters,* forthcoming (posted April 2, 2017), https://ssrn.com/abstract=2944919.

55 **"under $1,000, which is code for $999":** Matthew Amster-Burton, "Price Anchoring, or Why a $499 iPad Seems Inexpensive," *Mint-Life,* April 6, 2010, https://blog.mint.com/how-to/price-anchoring.

55 **sellers often use price to deliberately obscure information:** Authors' conversation with Florian Bauer, December 19, 2016.

CHAPTER 4: DATA-RICH MARKETS

59 **"Nothing anyone does will seem that crazy":** Olivia Solon, "Oh the Humanity! Poker Computer Trounces Humans in Big Step for AI," *Guardian,* January 30, 2017, https://www.theguardian.com/technology/2017/jan/30/libratus-poker-artificial-intelligence-professional-human-players-competition.

59 **playing poker against Libratus:** Quoted in Ben Popper, "This AI Will Battle Poker Pros for $200,000 in Prizes," *Verge,* January 4, 2017, http://www.theverge.com/2017/1/4/14161080/ai-vs-humans-poker-cmu-libratus-no-limit-texas-hold-em.

59 **Libratus remained cool:** Michael Laakasuo, Jussi Palomäki, and Mikko Salmela, "Experienced Poker Players Are Emotionally Stable," *Cyberpsychology, Behavior, and Social Networking* 17, no. 10 (October 2014), 668–671, http://online.liebertpub.com/doi/abs/10.1089/cyber.2014.0147.

60 **"plugging its own holes every night":** Authors' conversation with Jason Les, February 7, 2017.

60 **"we would call him a machine":** Ibid.

. 61 **Libratus racked up more than $1.7 million:** Solon, "Oh the Humanity!"

62 **the average individual wins weren't spectacular:** For a detailed description of Libratus's winning approach, see Nikolai Yakovenko, "CMU's Libratus Bluffs Its Way to Victory in #BrainsVsAI Poker Match," *Medium*, February 1, 2017, https://medium.com/@Moscow 25/cmus-libratus-bluffs-its-way-to-victory-in-brainsvsai-poker -match-99abd31b9cd4; see also Noam Brown and Tuomas Sandholm, "Safe and Nested Endgame Solving for Imperfect-Information Games" (2016), *Proceedings of the AAAI-17 Workshop on Computer Poker and Imperfect Information Games*, http:// www.cs.cmu.edu/~noamb/papers/17-AAAI-Refinement.pdf.

65 **get matched along multiple dimensions:** "Ride-Sharing with BlaBlaCar's New MariaDB Databases," ComparetheCloud.net, February 19, 2016, https://www.comparethecloud.net/articles /ride-sharing-with-blablacars-new-mariadb-databases; "About Us," BlaBlaCar.com, accessed January 27, 2017, https://www.blablacar .com/about-us.

65 **4 million people book rides:** Arun Sundararajan, "Uber and Airbnb Could Reverse America's Decades-Long Slide into Mass Cynicism," *Quartz*, June 9, 2016, https://qz.com/700859/uber-and -airbnb-will-save-us-from-our-decades-long-slide-into-mass -cynicism.

69 **Madi Solomon, an expert in such data:** Madi Solomon, "Transformational Metadata and the Future of Content Management: An Interview with Madi Solomon of Pearson PLC," *Journal of Digital Asset Management* 5, no. 1, 27–37, http://link.springer.com/article /10.1057/dam.2008.48; quote from conversation with Viktor Mayer-Schönberger.

70 **to automatically categorize product information:** Chris Mellor, "Metadata Manipulation by Alation Seeks Needles in Data Haystack," *Register*, April 1, 2015, http://www.theregister.co .uk/2015/04/01/metadata_manipulation_by_alation. See also Laura Melchior, "So stellt sich eBay im Bereich Daten und KI auf," *Internet World*, January 23, 2017, http://www.internetworld.de

/e-commerce/ebay/so-stellt-ebay-im-bereich-daten-ki-1188619
.html.

72 **these algorithms are the method by which:** There is a diverse
and growing literature on matching algorithms and processes; for
a good discussion of the state of research and how far it has come,
see Marzena Rostek and Nathan Yoder, "Matching with Multilat-
eral Contracts" (July 2, 2017), available at SSRN: https://ssrn.com
/abstract=2997223.

73 **carefully orchestrating the matching process:** See, e.g., Yash
Kanoria and Daniela Saban, "Facilitating the Search for Partners
on Matching Platforms: Restricting Agents' Actions" (July 5, 2017),
available at SSRN: https://ssrn.com/abstract=3004814.

73 **clearinghouse often collects preference information:** Alvin E.
Roth and Elliott Peranson, "The Redesign of the Matching Market
for American Physicians: Some Engineering Aspects of Economic
Design," *American Economic Review* 89, no. 4 (September 1999),
748–780.

74 **two of the world's leading experts in matching:** Alvin E. Roth,
*Who Gets What—and Why: The New Economics of Matchmaking
and Market Design* (New York: Houghton Mifflin Harcourt, 2015);
see also David S. Evans and Richard Schmalensee, *Matchmakers:
The New Economics of Multisided Platforms* (Cambridge: Harvard
Business Review Press, 2016).

75 **algorithm predicted which team would win:** Tim Adams, "Job
Hunting Is a Matter of Big Data, Not How You Perform at an
Interview," *Observer*, May 10, 2014, https://www.theguardian
.com/technology/2014/may/10/job-hunting-big-data-interview
-algorithms-employees; Sue Tabbitt, "Forget Myers-Briggs: Algorithms Can Better Predict Team Chemistry," *Guardian*, May
27, 2016, https://www.theguardian.com/small-business-network
/2016/may/27/forget-myers-briggs-algorithms-predict-team
-chemistry.

75 **Shepherd has replicated those results:** Oscar Williams-Grut,
"This Startup Can Predict If Your Business Will Fail with Questions Like 'Do You Like Horror Films?'" *Business Insider*, December
16, 2015, http://uk.businessinsider.com/simple-questions-like-do
-you-like-horror-films-can-predict-whether-a-startup-will-implode
-2015-12.

77   **representative of Big Data:** Viktor Mayer-Schönberger and Kenneth N. Cukier, *Big Data: A Revolution That Will Transform How We Live, Work, and Think* (New York: Houghton Mifflin Harcourt, 2013).

78   **going beyond its initial training:** For those who want to learn more about machine-learning methods (rather than Big Data more generally) in an easily accessible way, see Ethem Alpaydin, *Machine Learning* (Cambridge: MIT Press, 2016).

78   **Tesla's semiautonomous driving system:** Dana Hull, "The Tesla Advantage: 1.3 Billion Miles of Data," *Bloomberg Technology*, December 20, 2016, https://www.bloomberg.com/news/articles/2016-12-20/the-tesla-advantage-1-3-billion-miles-of-data.

83   **"supermarkets of love":** Julia M. Klein, "When Dating Algorithms Can Watch You Blush," *Nautilus*, April 14, 2016, http://nautil.us/issue/35/boundaries/when-dating-algorithms-can-watch-you-blush.

84   **it used the wrong data:** See, e.g., Paul W. Eastwick, Laura B. Luchies, Eli F. Finkel, and Lucy L. Hunt, "The Predictive Validity of Ideal Partner Preferences: A Review and Meta-Analysis," *Psychological Bulletin* 140 (20014), 623–665.

CHAPTER 5: COMPANIES AND CONTROL

87   **annual revenues of more than $100 billion:** Jim Milliot, "Amazon Sales Top $100 Billion," *Publishers Weekly*, January 28, 2016, http://www.publishersweekly.com/pw/by-topic/industry-news/financial-reporting/article/69269-amazon-sales-top-100-billion.html.

88   **"ordinary control freaks look like stoned hippies":** Steve Yegge's Google Plus post is archived, with his apparent permission, at https://plus.google.com/+RipRowan/posts/eVeouesvaVX.

88   **demands placed on employees:** Gregory Ferenstein, "Is Working at Amazon Terrible? According to Public Data, It's the Same as Much of Silicon Valley," *Forbes*, August 17, 2015, http://www.forbes.com/sites/gregoryferenstein/2015/08/17/is-working-at-amazon-terrible-according-to-public-data-its-the-same-as-much-of-silicon-valley/#5b68ce4a5f89.

88   **they have no autonomy:** See, for example, these reviews: https://
     www.glassdoor.co.uk/Reviews/Employee-Review-Amazon-com
     -RVW10200125.htm.

88   **"held accountable for a staggering array":** Jodi Kantor and David
     Streitfeld, "Inside Amazon: Wrestling Big Ideas in a Bruising Work-
     place," *New York Times*, August 15, 2015, https://www.nytimes
     .com/2015/08/16/technology/inside-amazon-wrestling-big
     -ideas-in-a-bruising-workplace.html.

89   **"you become an Amabot":** Ibid.

89   **get a warning or get fired:** Martha C. White, "Amazon's Use of
     'Stack' Ranking for Workers May Backfire, Experts Say," NBC
     News, August 17, 2015, http://www.nbcnews.com/business
     /business-news/amazons-use-stack-ranking-workers-may-backfire
     -experts-say-n411306.

89   **after the scientific management principles:** "Digital Taylorism,"
     Economist, September 12, 2015, http://www.economist.com/news
     /business/21664190-modern-version-scientific-management
     -threatens-dehumanise-workplace-digital.

90   **The firm can be many things:** See, e.g., John Micklethwait and
     Adrian Wooldridge, *The Company: A Short History of a Revolution-
     ary Idea* (New York: Modern Library, 2003).

91   **widespread adoption of Arabic numerals:** Alfred W. Crosby, *The
     Measure of Reality: Quantification and Western Society, 1250–1600*
     (Cambridge, UK: Cambridge University Press, 1997), 49.

91   **preeminent bankers of fifteenth-century Europe:** Jacob Soll, *The
     Reckoning: Financial Accountability and the Rise and Fall of Nations*
     (New York: Basic Books, 2014), 29–47.

92   **merchants in Florence were required to maintain:** Ibid., 35.

92   **annual audit, conducted by Cosimo himself:** Ibid., 37–38.

93   **trying to embezzle monies:** See Crosby, *The Measure of Reality*,
     204; Soll, *The Reckoning*, 37–38.

94   **Wedgwood transformed the information flows:** Soll, *The Reckon-
     ing*, 117–131.

95   **when bookkeepers get creative:** "Creative" accounting has been
     associated with spectacular bankruptcies and scandals, rang-
     ing from those befalling National City Bank (now Citibank, see
     "Stock Exchange Practices: Report of the Committee on Banking

and Currency" [the Pecora Commission Report], 73rd Congress, 2nd Session, report no. 1455, June 6, 1934, https://www.senate .gov/artandhistory/history/common/investigations/pdf/Pecora _FinalReport.pdf; for further details, see http://www.senate.gov /artandhistory/history/common/investigations/Pecora.htm) and drug company McKesson & Robbins (which faked purchase orders and inflated inventory, see Michael Chatfield, "McKesson & Robbins Case," in Michael Chatfield and Richard Vangermeersch, eds., *History of Accounting: An International Encyclopedia* [New York: Garland Publishing, 1996], 409–410) in the twentieth century to a string of high-profile cases at the turn of the millennium, including those of WorldCom (see Justin Kuepper, "Spotting Creative Accounting on the Balance Sheet," *Forbes*, March 25, 2010, http://www.forbes.com /2010/03/25/balance-sheet-tricks-personal-finance-accounting .html), Chiquita Brands, HealthSouth (see Michael J. Jones, *Creative Accounting, Fraud and International Accounting Scandals* [Chichester, England: John Wiley, 2011]), Enron (David Teather, "Billions Still Hidden in Enron Pyramid," *Guardian*, January 30, 2002, https:// www.theguardian.com/business/2002/jan/30/corporatefraud .enron2; Malcolm S. Salter, "Innovation Corrupted: The Rise and Fall of Enron (A)," Harvard Business School case study 905-048, December 2004 [revised October 2005], http://www.hbs.edu /faculty/Pages/item.aspx?num=31813), Lehman Brothers (see Rosalind Z. Wiggins and Andrew Metrick, "The Lehman Brothers Bankruptcy C: Managing the Balance Sheet Through the Use of Repo 105," Yale Program on Financial Stability case study 2014-3C-V1, October 1, 2014, https://papers.ssrn.com/sol3/papers .cfm?abstract_id=2593079; Donald J. Smith, "Hidden Debt: From Enron's Commodity Prepays to Lehman's Repo 105s," *Financial Analysts Journal* 67, no. 5 [September/October 2011], https://www .cfainstitute.org/learning/products/publications/faj/Pages/faj.v67 .n5.2.aspx), and electronics giant Toshiba, which in 2015 was caught posting profits early and pushing back the posting of losses in a salesperson's version of a Ponzi scheme (see Sean Farrell, "Toshiba Boss Quits over £780 Million Accounting Scandal," *Guardian*, July 21, 2015, https://www.theguardian.com/world/2015/jul/21/toshiba -boss-quits-hisao-tanaka-accounting-scandal).

95   **collection of minute details about every task:** Robert Kanigel, *The*

One Best Way: Frederick Winslow Taylor and the Enigma of Efficiency (New York: Little Brown, 1997).

96   **the first master's degree in business administration:** Soll, *The Reckoning*, 187.

96   **the punch-card tabulator:** Geoffrey D. Austrian, *Herman Hollerith: Forgotten Giant of Information Processing* (New York: Columbia University Press, 1982), 111 et seq. (chap. 9).

98   **other than through Billy Durant himself:** David A. Garvin and Lynne C. Levesque, "Executive Decision Making at General Motors," Harvard Business School case study 9-305-026, February 14, 2006, 2, http://www.hbs.edu/faculty/Pages/item.aspx?num=31870.

98   **leaving the company paralyzed:** William Pelfrey, *Billy, Alfred, and General Motors: The Story of Two Unique Men, a Legendary Company, and a Remarkable Time in American History* (New York: Amacom, 2006), 226.

99   **he hired an outside firm to collect the data:** Ibid., 260.

99   **could make decisions with up-to-date information:** John T. Landry, "Did Professional Management Cause the Fall of GM?" *Harvard Business Review*, June 9, 2009, https://hbr.org/2009/06/professional-management-and-th.

99   **hiring a series of data "whiz kids":** Phil Rosenzweig, "Robert S. McNamara and the Evolution of Modern Management," *Harvard Business Review*, December 2010, https://hbr.org/2010/12/robert-s-mcnamara-and-the-evolution-of-modern-management.

99   **simplifying data to make it more digestible:** Mayer-Schönberger and Cukier, *Big Data*, 164–165, 168.

100   **tools to shape the flow of information:** Ludwig Siegele and Joachim Zepelin, *Matrix der Welt: SAP und der neue globale Kapitalismus* (Frankfurt: Campus Verlag, 2009).

100   **the "noise" they create in the organization:** Daniel Kahneman, Andrew M. Rosenfield, Linnea Gandhi, and Tom Blaser, "Noise: How to Overcome the High, Hidden Cost of Inconsistent Decision Making," *Harvard Business Review* (October 2016), https://hbr.org/2016/10/noise.

101   **Checklists in aircraft:** Brigette M. Hales and Peter J. Pronovost, "The Checklist—a Tool for Error Management and Performance," *Journal of Critical Care* 21 (2006), 231–235.

101 **checklist in a hospital intensive care unit:** Atul Gawande, *The Checklist Manifesto: How to Get Things Right* (New York: Metropolitan Books, 2009). Gawande was inspired to test the checklist approach after reading about a pilot study conducted by Peter Pronovost of the Johns Hopkins University School of Medicine.

101 **balancing centralized and delegated decision:** Yingyi Qian, Gérard Roland, and Chenggang Xu, "Coordinating Changes in M-Form and U-Form Organizations," paper presented to the Nobel Symposium, April 1998, https://papers.ssrn.com/sol3/papers .cfm?abstract_id=163108.

101 **"decentralization with coordinated control":** Alfred P. Sloan Jr., *My Years with General Motors* (New York: Doubleday, 1990), 129, quoted in Garvin and Levesque, "Executive Decision Making at General Motors."

102 **a range of fundamental cognitive limitations:** Amos Tversky and Daniel Kahneman, "Judgment Under Uncertainty: Heuristics and Biases," *Science* 185, no. 4157 (September 27, 1974), 1124–1131. In 2002, Kahneman earned the Nobel Prize in Economics for the Tversky-Kahneman research; Tversky died in 1996 and therefore did not share in the honor. See also Daniel Kahneman, *Thinking, Fast and Slow* (New York: Farrar, Straus and Giroux, 2011); on how Kahneman and Tversky achieved their breakthrough insights, see Michael Lewis, *The Undoing Project: A Friendship That Changed Our Minds* (New York: W. W. Norton, 2016).

103 *confirmation bias:* Yoram Bar-Tal and Maria Jarymowicz, "The Effect of Gender on Cognitive Structuring: Who Are More Biased, Men or Women?" *Psychology* 1, no. 2 (January 2010), 80–87, http:// www.scirp.org/journal/PaperInformation.aspx?paperID=2096.

103 *fundamental attribution error:* Incheol Choi and Richard E. Nisbett, "Situational Salience and Cultural Differences in the Correspondence Bias and Actor-Observer Bias," *Personality and Social Psychology Bulletin* 24, no. 9 (September 1998), 949–960, http:// journals.sagepub.com/doi/abs/10.1177/0146167298249003; Minas N. Kastanakis and Benjamin G. Voyer, "The Effect of Culture on Perception and Cognition: A Conceptual Framework," *Journal of Business Research* 67, no. 4 (April 2014), 425–433, http://eprints.lse.ac.uk/50048/1/__lse.ac.uk_storage_LIBRARY _Secondary_libfile_shared_repository_Content_Voyer,%20B

_Effect%20culture%20perception_Voyer_Effect%20culture%20 perception_2014.pdf.

104 **limits of our ability to make optimal decisions:** Herbert A. Simon, *Models of Bounded Rationality* (Cambridge: MIT Press, 1982).

105 **"dangers of possessing too much information":** Gerd Gigerenzer, *Gut Feelings: The Intelligence of the Unconscious* (New York: Viking, 2007), 38.

105 **informational ignorance works just as well:** Cultivating "a beneficial degree of ignorance" is one of the self-help leitmotifs woven through Gerd Gigerenzer's international best seller *Gut Feelings*. The wisdom of gut instincts and intuition was also featured in Malcolm Gladwell's *Blink: The Power of Thinking Without Thinking* (New York: Little, Brown, 2005) and, more recently, in economists Donald Sull and Kathleen Eisenhardt's *Simple Rules: How to Thrive in a Complex World* (New York: Houghton Mifflin Harcourt, 2015).

105 **they value their gut feelings:** *Economist* Intelligence Unit, "Decisive Action: How Businesses Make Decisions and How They Could Do It Better" (London: Applied Predictive Technologies, n.d.), 7, http:// www.datascienceassn.org/sites/default/files/Decisive%20Action %20-%20How%20Businesses%20Make%20Decisions%20and %20How%20They%20Could%20do%20it%20Better.pdf.

106 **every second-best solution to a problem:** See generally David G. Myers, *Intuition—Its Powers and Its Perils* (New Haven: Yale University Press, 2002).

Chapter 6: Firm Futures

109 **company announced it would use Watson:** Justin McCurry, "Japanese Company Replaces Office Workers with Artificial Intelligence," *Guardian*, January 5, 2017, https://www.theguardian.com /technology/2017/jan/05/japanese-company-replaces-office -workers-artificial-intelligence-ai-fukoku-mutual-life-insurance; also see Fukoku Mutual's press release: https://translate.google.com /translate?depth=1&hl=en&prev=search&rurl=translate.google .com&sl=ja&sp=nmt4&u=http://www.fukoku-life.co.jp/about /news/download/20161226.pdf.

110 **20 percent of Daimler's global workforce:** "Daimler baut Konzern für die Digitalisierung um," *Frankfurter Allgemeine Zeitung,*

September 7, 2016, www.faz.net/aktuell/wirtschaft/daimler-baut
-konzern-fuer-die-digitalisierung-um-14424858.html.

110   **"supplement the hierarchical-management pyramid":** "Daimler
Chief Plots Cultural Revolution," *Handelsblatt Global,* July 25, 2016,
https://global.handelsblatt.com/companies-markets/daimlerchief
-plots-cultural-revolution-574783.

112   **required considerable institutional support:** See, e.g., Douglas W.
Allen, *The Institutional Revolution—Measurement and the Economic
Emergence of the Modern World* (Chicago: University of Chicago
Press, 2012).

114   **bump artificial intelligence into the executive ranks:** See also
Vegard Kolbjørnsrud, Richard Amico, and Robert J. Thomas, "How
Artificial Intelligence Will Redefine Management," *Harvard Business Review,* November 2, 2016, https://hbr.org/2016/11/how
-artificial-intelligence-will-redefine-management.

114   **automate three-fourths of all management decisions:** Olivia
Solon, "World's Largest Hedge Fund to Replace Managers with
Artificial Intelligence," *Guardian,* December 22, 2016, https://
www.theguardian.com/technology/2016/dec/22/bridgewater
-associates-ai-artificial-intelligence-management.

115   **making decisions less biased:** See Maarten Goos and Alan Manning,
"Lousy and Lovely Jobs: The Rising Polarization of Work in Britain,"
Review of Economics and Statistics 89 (February 2007), 118–133.

115   **learning decision-assistance system . . . construction company:** This case illustration is based on Emmanuel Marot, "Robot CEO: Your Next Boss Could Run on Code," *Venture Beat,*
March 20, 2016, https://venturebeat.com/2016/03/20/robot-ceo
-your-next-boss-could-run-on-code.

118   **T-shaped skill set:** Tom Kelley, *The Ten Faces of Innovation* (New
York: Doubleday, 2005), 75–78; on what skills are needed see also
Richard Susskind and Daniel Susskind, *The Future of Professions*
(Oxford: Oxford University Press, 2015).

121   **Spotify, a recent digital darling:** The examination of Spotify
is based on Thomas Ramge's visit to Spotify's HQ in Stockholm
on January 20–21, 2015; his analysis was published as Thomas
Ramge, "Nicht fragen. Machen," *brand eins* 033/15, https://www
.brandeins.de/archiv/2015/fuehrung/spotify-nicht-fragen
-machen/; for an English-language description of Spotify's unique

organizational setup, see Michael Mankins and Eric Garton, "How Spotify Balances Employee Autonomy and Accountability," *Harvard Business Review*, February 9, 2017, https://hbr.org/2017/02/how-spotify-balances-employee-autonomy-and-accountability.

123 **Ek has been described as:** Brendan Greeley, "Daniel Ek's Spotify: Music's Last Best Hope," *BloombergBusinessweek*, July 14, 2011, https://www.bloomberg.com/news/articles/2011-07-13/daniel-ek-s-spotify-music-s-last-best-hope.

124 **tool to elicit frequent feedback from others:** See Mankins and Garton, "How Spotify Balances Employee Autonomy and Accountability."

125 **John Deere's transition:** Darrell K. Rigby, Jeff Sutherland, and Hirotaka Takeuchi, "Embracing Agile," *Harvard Business Review*, May 2016, https://hbr.org/2016/05/embracing-agile.

125 **General Electric and Siemens are decentralizing:** "The Multinational Company Is in Trouble," *Economist*, January 28, 2017, http://www.economist.com/news/leaders/21715660-global-firms-are-surprisingly-vulnerable-attack-multinational-company-trouble.

125 **media giant Thomson Reuters aims to:** Mary Johnson, "How to Kickstart Innovation at a Multinational Corporation," Thomson Reuters blog, April 7, 2016, https://blogs.thomsonreuters.com/answerson/kickstart-innovation-multinational-corporation.

126 **recruit the talent their firms required:** Eben Harrell, "The Solution to the Skills Gap Could Already Be Inside Your Company," *Harvard Business Review*, September 27, 2016, https://hbr.org/2016/09/the-solution-to-the-skills-gap-could-already-be-inside-your-company.

127 **fluidity of human labor inside an organization:** Lowell L. Bryan, Claudia I. Joyce, and Leigh M. Weiss, "Making a Market in Talent," *McKinsey Quarterly*, May 2006, http://www.mckinsey.com/business-functions/organization/our-insights/making-a-market-in-talent.

128 **Procter & Gamble has even opened its internal platform:** John Horton, William R. Kerr, and Christopher Stanton, "Digital Labor Markets and Global Talent Flows," National Bureau of Economic Research Working Paper 23398 (May 2017), http://www.nber.org/papers/w23398.

CHAPTER 7: CAPITAL DECLINE

133 **"immeasurable tropical energy to create the perfect storm":** "NOAA Meteorologist Bob Case, the Man Who Named the Perfect Storm," *NOAA News*, June 16, 2000, http://www.noaanews .noaa.gov/stories/s444.htm.

133 **more than $200 million in damages:** National Climatic Data Center, "'Perfect Storm' Damage Summary," October 1991, http://www .ncdc.noaa.gov/oa/satellite/satelliteseye/cyclones/pfctstorm91 /pfctstdam.html.

134 **more than $8 trillion was lost:** Roger C. Altman, "The Great Crash, 2008: A Geopolitical Setback for the West," *Foreign Affairs*, January/February 2009, https://www.foreignaffairs.com/articles /united-states/2009-01-01/great-crash-2008.

135 **competition that has shrunk banks' margins:** Federal Reserve Bank of Saint Louis, "Net Interest Margin for All U.S. Banks," updated February 14, 2017, https://fred.stlouisfed.org/series/USNIM.

135 **situation is at least as bad in Europe:** Federal Reserve Bank of Saint Louis, "Bank's Net Interest Margin for Euro Area," updated August 17, 2016, https://fred.stlouisfed.org/series/DDEI01EZA156NWDB.

135 **one in five banks in Germany will earn:** Andreas Dombret, Yalin Gündüz, and Jörg Rocholl, "Will German Banks Earn Their Cost of Capital?" (2017), Bundesbank Discussion Paper No. 01/2017, https://ssrn.com/abstract=2910286.

136 **never recovered from the financial crisis of 2007:** U.S. Bureau of Labor Statistics, All Employees: Financial Activities: Commercial Banking (CEU5552211001), retrieved from FRED, Federal Reserve Bank of St. Louis; https://fred.stlouisfed.org/series /CEU5552211001, April 2, 2017.

136 **bank branches employed 212,000 fewer:** Valentina Romei, "Why Europe's Banks Will Never Be the Same Again," *Financial Times*, August 8, 2016, http://blogs.ft.com/ftdata/2016/08/08 /why-europes-banks-will-never-be-the-same-again.

136 **private banks in Switzerland vanished:** Oliver Suess and Jan-Henrik Foerster, "Commerzbank Plans Job Cuts in Biggest Overhaul Since Bailout," Bloomberg LP, September 29, 2016, http://www .bloomberg.com/news/articles/2016-09-29/commerzbank -shares-climb-on-report-of-10-000-job-cuts-pending.

136 **Commerzbank will cut one in five:** Matthew Allen, "One in Ten Swiss Private Banks Disappeared in 2015," *SwissInfo*, August 25, 2016, http://www.swissinfo.ch/eng/split-fortunes_one-in-10 -swiss-private-banks-disappeared-in-2015/42398770.

136 **UniCredit bank will close 26 percent:** Martin Arnold, "Uni-Credit Boss Wastes No Time in Tackling the Bank's Problems," *Financial Times*, December 13, 2016, https://www.ft.com/content /0ed769fc-c0a6-11e6-9bca-2b93a6856354.

140 **Second Payment Service Directive:** Directive (EU) 2015/2366 of the European Parliament and of the Council of 25 November 2015 on payment services in the internal market, OJ L 337, 23.12.2015, 35–127, http://eur-lex.europa.eu/legal-content/EN /TXT/?uri=CELEX:32015L2366; see also "New European Rules Will Open Retail Banking," *Economist*, March 23, 2017, http://www .economist.com/news/leaders/21719476-dangers-privacy-and -security-are-outweighed-benefits-new-european-rules-will-open.

142 **wide variety of signals . . . can be honest:** Alex Pentland, *Honest Signals* (Cambridge: MIT Press, 2008).

143 **raised money simply because it could:** Leslie Hook, "Venture Capital Funding in Start-Ups Surges to $100bn for Quarter," *Financial Times*, October 14, 2015, https://www.ft.com/content /e95f5c6e-7238-11e5-bdb1-e6e4767162cc.

143 **the number of listed companies:** Maureen Farrell, "America's Roster of Public Companies Is Shrinking Before Our Eyes," *Wall Street Journal*, January 6, 2017, https://www.wsj.com/articles/americas-roster -of-public-companies-is-shrinking-before-our-eyes-1483545879.

148 **No digital currency is capable of:** For more on blockchain, see Don Tapscott and Alex Tapscott, *The Blockchain Revolution: How the Technology Behind Bitcoin Is Changing Money, Business, and the World* (New York: Portfolio/Penguin Books, 2016).

149 **fintechs attracted investments exceeding $19 billion:** Andrew Meola, "The Fintech Report 2016: Financial Industry Trends and Investment," *Business Insider*, December 14, 2016, http://www .businessinsider.de/the-fintech-report-2016-financial-industry -trends-and-investment-2016-12?r=US&IR=T; KPMG, "The Pulse of Fintech: Global Analysis of Fintech Venture Funding," November 13, 2016, https://assets.kpmg.com/content/dam/kpmg/xx/pdf /2016/11/the-pulse-of-fintech-q3-report.pdf.

149   **a fintech bubble:** Alessandro Hatami, "After the Fintech Bubble—
the Winners and Losers," BankNXT, February 15, 2016, http://
banknxt.com/55760/fintech-bubble-winners-and-losers.

151   **saving its customers . . . an estimated $1.45 billion:** https://www
.sofi.com.

151   **bringing data-rich consumer credit scoring to China:** Jon Russell,
"Baidu Invests in ZestFinance to Develop Search-Powered Credit
Scoring for China," *TechCrunch*, July 17, 2016, https://techcrunch
.com/2016/07/17/baidu-invests-in-zestfinance-to-develop
-search-powered-credit-scoring-for-china.

152   **total market for peer-to-peer lending in China:** "In Fintech China
Shows the Way," *Economist*, February 25, 2017.

154   **A counter example suggests:** The following history of investment
banking is based on the perceptive Alan D. Morrison and William J.
Wilhelm, Jr. *Investment Banking: Institutions, Politics, and Law* (Oxford: Oxford University Press, 2007); an article version is Alan Morrison and William Wilhelm, "Investment Banking: Past, Present,
and Future," *Journal of Applied Corporate Finance* 19 (2007), 8–20.

156   **lacks the insight, based on information:** Albert Wenger, *World After Capital*, https://worldaftercapital.gitbooks.io/worldaftercapital
/content/part-two/Capital.html.

CHAPTER 8: FEEDBACK EFFECTS

159   **built-in risk of extreme failure:** See the official report of the accident investigation BEA, *Final Report—On the Accident on 1st
June 2009 to the Airbus A330-203 Registered F-GZCP Operated by
Air France Flight AF 447 Rio de Janeiro–Paris*, July 2012, https://
www.bea.aero/docspa/2009/f-cp090601.en/pdf/f-cp090601
.en.pdf; see also William Langewiesche, "The Human Factor," *Van-
ity Fair*, September 17, 2014, http://www.vanityfair.com/news
/business/2014/10/air-france-flight-447-crash; Tim Harford, "Crash:
How Computers Are Setting Us Up for Disaster," *Guardian*, October 11, 2016, https://www.theguardian.com/technology/2016
/oct/11/crash-how-computers-are-setting-us-up-disaster.

159   **the general theory of feedback:** See George Dyson, *Turing's Cathe-*

dral: The Origins of the Digital Universe (New York: Pantheon Books, 2012), 109–114.

159   **in choosing the term "cybernetics":** On the ambivalence of Norbert Wiener's work, see Flo Conway and Jim Siegelman, *Dark Hero of the Information Age—In Search of Norbert Wiener, the Father of Cybernetics* (New York: Basic Books, 2005).

160   **"their control over the rest of the human race":** Norbert Wiener, The Human Use of Human Beings (Boston: Da Capo Press, 1988), 247–250.

160   **Champagne fairs of the Middle Ages:** Ray Fisman and Tim Sullivan, *The Inner Lives of Markets: How People Shape Them—and They Shape Us* (New York: PublicAffairs, 2016).

161   **search requests . . . go to Google:** "Marktanteile der Suchmaschinen weltweit nach mobiler und stationärer Nutzung im März 2017," https://de.statista.com/statistik/daten/studie/222849/umfrage /marktanteile-der-suchmaschinen-weltweit.

161   **Amazon's share of . . . online retail revenues:** "Amazon Accounts for 43 Percent of US Online Retail Sales," *Business Insider,* February 2, 2017, http://www.businessinsider.de/amazon-accounts-for-43 -of-us-online-retail-sales-2017-2?r=US&IR=T.

161   **Facebook's almost 2 billion users:** "Leading Countries Based on Number of Facebook Users as of April 2016 (in Millions)," https:// www.statista.com/statistics/268136/top-15-countries-based -on-number-of-facebook-users.

161   **GoDaddy is the largest domain-name registrar:** Andrew Allemann, "GoDaddy Marches Toward $1 Billion," *DomainNameWire,* August 17, 2010, http://domainnamewire.com/2010/08/17/go -daddy-marches-toward-1-billion.

161   **WordPress dominates . . . Netflix rules:** According to W3Techs, WordPress is used by almost 60 percent of all websites with a known content management system and by almost 30 percent of all websites; https://w3techs.com/technologies/details/cm -wordpress/all/all; in early 2017, Netflix's share of the streaming market in the United States was about 75 percent; see Sara Perez, "Netflix Reaches 75 percent of US Streaming Service Viewers, but YouTube Is Catching Up," *TechCrunch,* April 20, 2017,

https://techcrunch.com/2017/04/10/netflix-reaches-75-of-u-s
-streaming-service-viewers-but-youtube-is-catching-up.

163 **service itself did not improve:** See also Claude S. Fischer, *America Calling: A Social History of the Telephone to 1940* (Berkeley and Los Angeles: University of California Press, 1994).

163 **this network effect:** Carl Shapiro and Hal R. Varian, *Information Rules: A Strategic Guide to the Network Economy* (Boston: Harvard Business Review Press, 1999), 173 et seq.

164 **business dynamism, driven by innovative disruptors:** Ryan A. Decker, John Haltiwanger, Ron S. Jarmin, and Javier Miranda, "Declining Dynamism, Allocative Efficiency, and the Productivity Slowdown," Board of Governors of the Federal Reserve System, Finance and Economics Discussion Series 2017-019, https://doi.org/10.17016/FEDS.2017.019.

165 **antitrust lawsuit against Microsoft:** See, e.g., Andrew I. Gavil and Harry Fist, *The Microsoft Antitrust Case* (Cambridge: MIT Press, 2014).

165 **antitrust case against Google in Europe:** See, e.g., Benjamin Edelmann, "Does Google Leverage Market Power Through Tying and Bundling?" *Journal of Competition Law and Economics* 11, no. 2 (2015), 365–400, https://doi.org/10.1093/joclec/nhv016.

166 **go beyond regulating anticompetitive behavior:** Ariel Ezrachi and Maurice E. Stucke, *Virtual Competition: The Promise and Perils of the Algorithm-Driven Economy* (Cambridge: Harvard University Press, 2016); see also Maurice Stucke and Allen Grunes, *Big Data and Competition Policy* (New York: Oxford University Press, 2016).

167 **"open up" their algorithms:** For a critical view of algorithmic transparency, see Joshua A. Kroll et al., "Accountable Algorithms," *University of Pennsylvania Law Review* 165 (2017), 633–705, https://www.pennlawreview.com/print/165-U-Pa-L-Rev-633.pdf.

167 **economists . . . offer an intriguing idea:** Jens Prüfer and Christoph Schottmüller, "Competing with Big Data," February 16, 2017, TILEC Discussion Paper 2017-006, available at http://dx.doi.org/10.2139/ssrn.2918726; this extends an idea originally suggested in Cédric Argenton and Jens Prüfer, "Search Engine Competition with Network Externalities," *Journal of Competition Law and Economics* 8 (2012), 73–105, https://pure.uvt.nl/portal/files/1373523/search_engines.pdf.

171 **Homogeneity of the systems we employ:** We see similar problems emerge with crops and fruits lacking genetic variety, such as the conventional banana (the "Cavendish") that is threatened by its vulnerability to a dangerous fungus; see, e.g., "A Future with No Bananas?" *New Scientist*, May 13, 2006, https://www.newscientist.com/article/dn9152-a-future-with-no-bananas.

173 **Critics of fair-value accounting:** For a discussion and critical analysis about the pros and cons of the argument, see Christian Laux and Christian Leuz, "Did Fair-Value Accounting Contribute to the Financial Crisis?" *Journal of Economic Perspectives* 24 (2010), 93–118.

174 **shift in focus from collection to use:** See, e.g., Fred H. Cate and Viktor Mayer-Schönberger, "Notice and Consent in a World of Big Data," *International Data Privacy Law* 3 (2013), 67–73; Kirsten E. Martin, "Transaction Costs, Privacy, and Trust: The Laudable Goals and Ultimate Failure of Notice and Choice to Respect Privacy Online," *First Monday* 18, no. 12-2 (2013), http://firstmonday.org/ojs/index.php/fm/article/view/4838/3802; Alessandro Mantelero, "The Future of Consumer Data Protection in the E.U. Rethinking the 'Notice and Consent' Paradigm in the New Era of Predictive Analytics," *Computer Law and Security Review* 30 (2014), 643; Joel R. Reidenberg et al., "Privacy Harms and the Effectiveness of the Notice and Choice Framework," *I/S* 11 (2015), 485–524, http://moritzlaw.osu.edu/students/groups/is/files/2016/02/10-Reidenberg-Russell-Callen-Qasir-and-Norton.pdf.

175 **computer system that would assist the Chilean government:** The evolution of Cybersyn and its implications are eloquently captured in Eden Medina, *Cybernetics Revolutionaries: Technology and Politics in Allende's Chile* (Cambridge: MIT Press, 2011); see also Evgeny Morozov, "The Planning Machine," *New Yorker*, October 13, 2014, http://www.newyorker.com/magazine/2014/10/13/planning-machine. The development of Cybersyn also underlies the plot of a work of fiction: see Sascha Reh, *Gegen die Zeit* (Frankfurt, Germany: Schöffling, 2015).

177 **Great Famine of 1932–1933:** See Anne Applebaum, *Red Famine: Stalin's War on Ukraine* (New York: Doubleday, 2017).

178 **to coax us to transact appropriately:** See, e.g., Richard H. Thaler and Cass R. Sunstein, *Nudge: Improving Decisions About Health, Wealth, and Happiness* (New Haven: Yale University Press, 2008).

CHAPTER 9: UNBUNDLING WORK

181   **"The drive was as mundane":** Alex Davies, "Uber's Self-Driving Truck Makes Its First Delivery: 50,000 Beers," *Wired*, October 25, 2016, https://www.wired.com/2016/10/ubers-self-driving-truck-makes-first-delivery-50000-beers.

182   **the system won't take away their jobs:** Eric Newcomer and Alex Webb, "Uber Self-Driving Truck Packed with Budweiser Makes First Delivery in Colorado," *Bloomberg*, October 25, 2016, https://www.bloomberg.com/news/articles/2016-10-25/uber-self-driving-truck-packed-with-budweiser-makes-first-delivery-in-colorado.

183   **median annual income of more than $40,000:** "Heavy and Tractor-Trailer Truck Drivers," Bureau of Labor Statistics, *Occupational Outlook Handbook*, https://www.bls.gov/ooh/transportation-and-material-moving/heavy-and-tractor-trailer-truck-drivers.htm.

183   **middle-income desk jobs that . . . will disappear:** Michael Chui, James Manyika, and Mehdi Miremadi, "Where Machines Could Replace Humans—and Where They Can't (Yet)," *McKinsey Quarterly* (July 2016), http://www.mckinsey.com/business-functions/digital-mckinsey/our-insights/where-machines-could-replace-humans-and-where-they-cant-yet.

183   **labor force has declined from its peak:** The participation rate in 2017 was around 63 percent, down from over 67 percent in 2000, and below the level it had been in more than three decades; see U.S. Bureau of Labor Statistics, Labor Force Participation Rates, data sets and graphs available at https://data.bls.gov.

184   **forecast depressing employment figures:** Erik Brynjolfsson and Andrew McAfee, *The Second Machine Age: Work, Progress, and Prosperity in a Time of Brilliant Technologies* (New York: W. W. Norton, 2016); Carl Benedikt Frey and Michael A. Osborne, *The Future of Employment: How Susceptible Are Jobs to Computerisation?* (Oxford, UK: Oxford Martin School, September 17, 2013), http://www.oxfordmartin.ox.ac.uk/downloads/academic/The_Future_of_Employment.pdf.

184   **advent of a "second machine age":** Brynjolfsson and McAfee, *The Second Machine Age*.

184   **"labor share" has declined considerably:** Matthias Kehrig and

Nicolas Vincent, "Growing Productivity Without Growing Wages: The Micro-Level Anatomy of the Aggregate Labor Share Decline," CESifo Working Paper Series No. 6454, May 3, 2017, https://ssrn .com/abstract=2977787.

184 **In most advanced economies, labor share:** International Labor Organization and Organisation for Economic Co-operation and Development, "The Labour Share in G20 Economies" (February 2015), 11, https://www.oecd.org/g20/topics/employment -and-social-policy/The-Labour-Share-in-G20-Economies.pdf; *OECD Employment Outlook 2012*, 115, http://www.oecd-ilibrary .org/employment/oecd-employment-outlook-2012_empl _outlook-2012-en.

184 **fallen in the large economies of India and China:** Loukas Karabarbounis and Brent Neiman, "The Global Decline of the Labor Share," *Quarterly Journal of Economics* 129, no. 1 (January 2013), 61–103, n1.

184 **labor share globally has been dropping:** Ibid, 1.

185 **something that affects them all:** *OECD Employment Outlook 2012*, 118–119.

185 **displaces blue-collar and low- and middle-income:** Ibid., 115–116.

186 **temporary gigs with limited or no benefits:** Ian Hathaway and Mark Muro, "Tracking the Gig Economy: New Numbers," *Brookings Institution*, October 13, 2016, https://www.brookings.edu /research/tracking-the-gig-economy-new-numbers; the gig economy is not limited to advanced economies; see, e.g., Mark Graham, Isis Hjorth, and Vili Lehdonvirta, "Digital Labour and Development: Impacts of Global Digital Labour Platforms and the Gig Economy on Worker Livelihoods," *Transfer: European Review of Labour and Research*, March 16, 2017, http://journals.sagepub.com /eprint/3FMTvCNPJ4SkhW9tgpWP/full.

186 **shrinking role of labor:** Thomas Piketty, *Capital in the Twenty-First Century* (Cambridge, MA: Belknap Press, 2014).

187 **not necessarily to tax the economy more:** See Ryan Abbott and Bret N. Bogenschneider, "Should Robots Pay Taxes? Tax Policy in the Age of Automation," forthcoming in *Harvard Law and Policy Review* (March 13, 2017), https://ssrn.com/abstract=2932483.

187 **Bill Gates, announced his support:** Kevin J. Delaney, "The Robot That Takes Your Job Should Pay Taxes, Says Bill Gates,"

Quartz, February 17, 2017, https://qz.com/911968/bill-gates-the
-robot-that-takes-your-job-should-pay-taxes.

187 **it might stifle innovation:** Georgina Prodhan, "European Parliament Calls for Robot Law, Rejects Robot Tax," Reuters, February 16, 2017, http://www.reuters.com/article/us-europe-robots
-lawmaking-idUSKBN15V2KM.

189 **enthusiasm for UBI:** For an excellent monograph on universal basic income, see Philippe van Parijs and Yannick Vanderborght, *Basic Income: A Radical Proposal for a Free Society and a Sane Economy* (Cambridge: Harvard University Press, 2017).

190 **Paine proposed a basic income:** Thomas Paine, *Agrarian Justice* (1797), https://www.ssa.gov/history/tpaine3.html.

190 **Friedman suggested a negative income tax:** Milton Friedman, *Capitalism and Freedom* (Chicago: University of Chicago Press, 1962).

190 **Democratic presidential candidate George McGovern:** Van Parijs and Vanderborght, *Basic Income*, 90–93.

190 **Nixon then proposed his own family-assistance program:** Peter Passell and Leonard Ross. "Daniel Moynihan and President-Elect Nixon: How Charity Didn't Begin at Home," *New York Times*, January 14, 1973, http://www.nytimes.com/books/98/10/04/specials
/moynihan-income.html.

191 **empower people to choose for themselves:** Nathan Schneider, "Why the Tech Elite Is Getting Behind Universal Basic Income," *Vice*, January 6, 2015, https://www.vice.com/en_au/article
/something-for-everyone-0000546-v22n1.

191 **effects of a UBI on human motivation:** Jon Henley, "Finland Trials Basic Income for Unemployed," *Guardian*, January 3, 2017, https://www.theguardian.com/world/2017/jan/03/finland
-trials-basic-income-for-unemployed.

191 **Switzerland held a national referendum:** Ibid.

191 **Canada experimented with a version:** Van Parijs and Vanderborght, *Basic Income*, 141–143; see also Zi-Ann Lum, "A Canadian City Once Eliminated Poverty and Nearly Everyone Forgot About It," *Huffington Post Canada*, December 23, 2014, http://www
.huffingtonpost.ca/2014/12/23/mincome-in-dauphin-manitoba
_n_6335682.html.

192 **subsidize the middle class and the affluent:** This is in contrast to
proposals such as Milton Friedman's negative income tax; Friedman,
Capitalism and Freedom.

192 **justify spending such a huge sum:** For more on how to calculate the
amount needed to fund a UBI and the income tax rate that would be necessary, see "Basically Unaffordable," *Economist,* May 23, 2015, http://
www.economist.com/news/finance-and-economics/21651897
-replacing-welfare-payments-basic-income-all-alluring.

194 **capital share evaporates:** Matthew Rognlie, "Deciphering the Fall
and Rise in the Net Capital Share: Accumulation or Scarcity?"
Brookings Papers on Economic Activity (Spring 2015), https://
www.brookings.edu/wp-content/uploads/2016/07/2015a
_rognlie.pdf.

194 **conventional calculations of capital share assume:** Simcha Barkai,
"Declining Labor and Capital Shares," http://home.uchicago
.edu/~barkai/doc/BarkaiDecliningLaborCapital.pdf.

194 **data processing requires less capital:** Loukas Karabarbounis and
Brent Neiman, "The Global Decline of the Labor Share," National
Bureau of Economic Research Working Paper 19136 (June 2013),
http://ww.nber.org/papers/w19136.pdf.

194 **labor benefits more from technology:** Robert Z. Lawrence,
"Recent Declines in Labor's Share in US Income: A Preliminary
Neoclassical Account," National Bureau of Economic Research
Working Paper 21296 (June 2015), http://www.nber.org/papers
/w21296.

195 **innovative activity and business dynamism:** Ryan A. Decker et al.,
"Declining Dynamism, Allocative Efficiency, and the Productivity
Slowdown," Finance and Economics Discussion Series 2017-019
(Washington, DC: Board of Governors of the Federal Reserve Sys-
tem, 2017), https://doi.org/10.17016/FEDS.2017.019.

195 **further light on the underlying dynamic:** David Autor et al.,
"The Fall of Labor Share and the Rise of Superstar Firms," National Bureau of Economic Research Working Paper 23396 (May
2017), http://www.nber.org/papers/w23396; David Autor et al.,
"Concentrating on the Fall of the Labor Share," National Bureau of
Economic Research Working Paper 23108 (January 2017), http://
www.nber.org/papers/w23108.

195 **they can achieve very high revenues:** This also applies to superstars among manufacturing plants, see Kehrig and Vincent, "Growing Productivity Without Growing Wages."

196 **five . . . are obvious superstars:** Dan Strumpf, "The Only Six Stocks That Matter," *Wall Street Journal*, July 26, 2015, https://www.wsj .com/articles/the-only-six-stocks-that-matter-1437942926.

197 **textbook example of a bad idea:** The exceptions would be social welfare programs that are "self-funded" through direct contributions from workers; in those cases, if policy makers reject the idea that corporate tax receipts should fund social welfare, a robo tax may be an option to make up for the shortfall caused by a decreasing labor share.

198 **over $200 billion in lost tax revenues:** Fabie Candau and Jacques Le Cacheux, "Corporate Income Tax as a Genuine Own Resource," March 23, 2017, https://ssrn.com/abstract=2939938.

198 **personal income would be taxed only:** See, e.g., Alan D. Viard and Robert Carroll, *Progressive Consumption Taxation: The X-Tax Revisited* (Washington, DC: American Enterprise Institute Press, 2012), but the idea is much older—see, e.g., William D. Andrews, "A Consumption-Type or Cash Flow Personal Income Tax," 87 *Harvard Law Review* 1113 (1974).

198 **support across the political spectrum:** In addition to politicians such as Democratic Senator Ben Cardin (https://www.cardin.senate .gov/pct) and think tanks such as the conservative American Enterprise Institute (AEI), advocates include high-tech figures such as Bill Gates (Tim Worstall, "Bill Gates Points to the Best Tax System, the Progressive Consumption Tax," *Forbes*, March 18, 2014, https://www.forbes.com/sites/timworstall/2014/03/18/bill -gates-points-to-the-best-tax-system-the-progressive -consumption-tax).

202 **to spot changes in skill demand:** We are not alone suggesting this; see, e.g., World Economic Forum, *The Future of Jobs Report* (January 2016), 24 and 29, http://www3.weforum.org/docs/WEF_Future _of_Jobs.pdf.

203 **"look to build a monopoly":** Peter Thiel, "Competition Is for Losers," *Wall Street Journal*, September 12, 2014, https://www.wsj.com /articles/peter-thiel-competition-is-for-losers-1410535536.

205 **choose the job they like:** See also Van Parijs and Vanderborght, *Basic Income*, 165–169.

CHAPTER 10: HUMAN CHOICE

207 **become "a CEO of a retail company":** Ryan Mac, "Stitch Fix: The $250 Million Startup Playing Fashionista Moneyball," *Forbes*, June 1, 2016, www.forbes.com/sites/ryanmac/2016/06/01/fashionista -moneyball-stitch-fix-katrina-lake/#54e798e859a2.

208 **"constantly maxing out Lake's $6,000-limit credit card":** Ibid.

209 **a potential start-up "unicorn":** "Fifty Companies That May Be the Next Start-Up Unicorns," *New York Times*, August 23, 2015, https://bits.blogs.nytimes.com/2015/08/23/here-are-the -companies-that-may-be-the-next-50-start-up-unicorns/?_r=0.

210 **enables Stitch Fix to be a matchmaker:** http://algorithms-tour .stitchfix.com.

211 **a social debt that most customers choose to repay:** Much like the personal debt that transaction partners have been found to want to repay when they receive positive feedback; see, e.g., Gary Bolton, Ben Greiner, and Alex Ockenfels, "Engineering Trust—Reciprocity in the Production of Reputation Information," *Management Science* 59, no. 2 (2013), 265–285.

213 **one in four trucks drive empty:** Robert Matthams, "Despite High Fuel Prices, Many Trucks Run Empty," *Christian Science Monitor*, February 25, 2012, http://www.csmonitor.com/Business/2012/0225 /Despite-high-fuel-prices-many-trucks-run-empty.

216 **one of the pioneer venture capitalists:** "Eugene Kleiner," Kleiner, Perkins, Caufield, and Byers, http://www.kpcb.com/partner/eugene -kleiner.

219 **fear is that as we delegate decisions to machines:** See, e.g., Nick Bostrom, *Superintelligence: Paths, Dangers, Strategies* (Oxford: Oxford University Press, 2014).

220 **The end of scarcity:** Ibid.

220 **"A world of increasing abundance":** Ibid.

221 **"Cartier for everyone":** Aaron Bastani, "Britain Doesn't Need More Austerity, It Needs Luxury Communism," *Vice*, June 12, 2015, https://www.vice.com/en_au/article/luxury-communism-933.

221 **"cybernetic meadow, tended to by machines":** Brian Merchant, "Fully Automated Luxury Communism," *Guardian*, March 18, 2015, https://www.theguardian.com/sustainable-business/2015/mar /18/fully-automated-luxury-communism-robots-employment.

223 **open the window to new insights:** Avi Loeb, "Good Data Are Not Enough," *Nature*, November 2, 2016, http://www.nature.com /news/good-data-are-not-enough-1.20906.

INDEX

VIKTOR MAYER-SCHÖNBERGER (left) is a professor at the University of Oxford and the coauthor, with Kenneth Cukier, of the best-selling *Big Data*.

THOMAS RAMGE is the technology correspondent of the business magazine *brand eins* and writes for the *Economist*.